Haloacetic Acid Removal Using Granular Activated Carbon

Awwa Research Foundation

About the Awwa Research Foundation

The Awwa Research Foundation (AwwaRF) is a member-supported, international, nonprofit organization that sponsors research to enable water utilities, public health agencies, and other professionals to provide safe and affordable drinking water to consumers.

The Foundation's mission is to advance the science of water to improve the quality of life. To achieve this mission, the Foundation sponsors studies on all aspects of drinking water, including supply and resources, treatment, monitoring and analysis, distribution, management, and health effects. Funding for research is provided primarily by subscription payments from approximately 1,000 utilities, consulting firms, and manufacturers in North America and abroad. Additional funding comes from collaborative partnerships with other national and international organizations, allowing for resources to be leveraged, expertise to be shared, and broad-based knowledge to be developed and disseminated. Government funding serves as a third source of research dollars.

From its headquarters in Denver, Colorado, the Foundation's staff directs and supports the efforts of more than 800 volunteers who serve on the board of trustees and various committees. These volunteers represent many facets of the water industry, and contribute their expertise to select and monitor research studies that benefit the entire drinking water community.

The results of research are disseminated through a number of channels, including reports, the Web site, conferences, and periodicals.

For subscribers, the Foundation serves as a cooperative program in which water suppliers unite to pool their resources. By applying Foundation research findings, these water suppliers can save substantial costs and stay on the leading edge of drinking water science and technology. Since its inception, AwwaRF has supplied the water community with more than $300 million in applied research.

More information about the Foundation and how to become a subscriber is available on the Web at **www.awwarf.org**.

Haloacetic Acid Removal Using Granular Activated Carbon

Prepared by:
Yuefeng F. Xie and **Hongwei Wu**
Penn State Harrisburg, Environmental Engineering Program
777 West Harrisburg Pike, Middletown, PA 17057

and

Hsin-hsin Tung
Penn State University Park, Department of Civil and Environmental Engineering
University Park, PA 16802

Sponsored by:
Awwa Research Foundation
6666 West Quincy Avenue, Denver, CO 80235-3098

Published by:

Awwa Research Foundation
Advancing the Science of Water®

IWA Publishing

DISCLAIMER

This study was funded by the Awwa Research Foundation (AwwaRF). AwwaRF assumes no responsibility for the content of the research study reported in this publication or for the opinions or statements of fact expressed in the report. The mention of trade names for commercial products does not represent or imply the approval or endorsement of AwwaRF. This report is presented solely for informational purposes.

Published by IWA Publishing, Alliance House, 12 Caxton Street, London SW1H 0QS, UK

Telephone: +44 (0) 20 7654 5500; Fax: +44 (0) 20 7654 5555; Email: publications@iwap.co.uk;
Web: **www.iwapublishing.com**

AwwaRF report number 91041F. Originally published by AwwaRF for its subscribers in 2004.
IWA Publishing version published 2005.

ISBN 1 84339 929 6

CONTENTS

TABLES

FIGURES

FOREWORD

The Awwa Research Foundation is a nonprofit corporation that is dedicated to the implementation of a research effort to help utilities respond to regulatory requirements and traditional high-priority concerns of the industry. The research agenda is developed through a process of consultation with subscribers and drinking water professionals. Under the umbrella of a Strategic Research Plan, the Research Advisory Council prioritizes the suggested projects based upon current and future needs, applicability, and past work; the recommendations are forwarded to the Board of Trustees for final selection. The foundation also sponsors research projects through an unsolicited proposal process; the Collaborative Research, Research Applications, and Tailored Collaboration programs; and various joint research efforts with organizations such as the U.S. Environmental Protection Agency, the U.S. Bureau of Reclamation, and the Association of California Water Agencies.

This publication is a result of one of these sponsored studies, and it is hoped that its findings will be applied in communities throughout the world. The following report serves not only as a means of communicating the results of the water industry's centralized research program but also as a tool to enlist the further support of the nonmember utilities and individuals.

Projects are managed closely from their inception to the final report by the foundation's staff and large cadre of volunteers who willingly contribute their time and expertise. The foundation serves a planning and management function and awards contracts to other institutions such as water utilities, universities, and engineering firms. The funding for this research effort comes primarily from the Subscription Program, through which water utilities subscribe to the research program and make an annual payment proportionate to the volume of water they deliver and consultants and manufacturers subscribe based on their annual billings. The program offers a cost-effective and fair method for funding research in the public interest.

A broad spectrum of water supply issues is addressed by the foundation's research agenda: resources, treatment and operations, distribution and storage, water quality and analysis, toxicology, economics, and management. The ultimate purpose of the coordinated effort is to assist water suppliers to provide the highest possible quality of water economically and reliably. The true benefits are realized when the results are implemented at the utility level. The foundation's trustees are pleased to offer this publication as a contribution toward that end.

Walter J. Bishop
Chair, Board of Trustees
Awwa Research Foundation

James F. Manwaring, P.E.
Executive Director
Awwa Research Foundation

ACKNOWLEDGMENTS

The authors of this report wish to acknowledge the following water utilities and individuals for their cooperation and assistance in this project:

- City of Harrisburg, Harrisburg, Pa., Gregg Haney and Daniel E. Standish
- Contra Costa Water District, Concord, Calif., Peter Zhou, David Huey, Pat Panus, Larry McCollum, Jean Zacher, and Paul Prewitt
- Calgon Carbon Corporation, Pittsburgh, Pa., Neal E. Megonnell
- Pennsylvania American Water Company, Hershey, Pa., Gary D. Witmeyer, Donald Holbrook, Fred J. Wendekier, and Sharon L. Fillmann
- United Water Pennsylvania, Harrisburg, Pa., Arthur C. Saunders

In addition, the authors would like to thank Richard F. Unz and Katherine H. Baker at the Pennsylvania State University for their technical assistance on this project. The financial support provided by the Pennsylvania State University Institutes of Environment is greatly appreciated.

The authors would also like to thank Awwa Research Foundation (AwwaRF) Project Advisory Committee members, Mary Jo Kirisits, Rengao Song, Roy L. Williams, and Sherry L. Williams, for their critical reviews and suggestions. Last, but not least, without the technical support of AwwaRF project officers, My-Linh Nguyen and Kenan Ozekin, this project would not have been possible.

EXECUTIVE SUMMARY

Haloacetic acids (HAAs) are a group of organic disinfection by-products (DBPs) regulated under the USEPA's Stage 1 Disinfectants/Disinfection By-Products Rule. The maximum contaminant level (MCL) for the sum of five haloacetic acids (HAA5) under the rule is 60 µg/L. These five HAAs are monochloroacetic acid (ClAA), monobromoacetic acid (BrAA), dichloroacetic acid (Cl_2AA), dibromoacetic acid (Br_2AA), and trichloroacetic acid (Cl_3AA). Enhanced coagulation/softening for precursor removal and alternative disinfectants are common means for controlling HAAs and other DBPs in finished water.

It is well documented that biological filtration effectively removes organic ozonation by-products and some natural organic matter. However, information is limited on the use of biofiltration for HAA removal. In 1993, Singer et al. observed a rapid elimination of HAAs in treated water during aquifer storage. A biological degradation mechanism was suggested. In 1994, Williams, Williams, and Rindfleisch reported extremely low levels of HAAs at the maximum residence time locations in three distribution systems in Virginia. Biological degradation of HAAs in distribution systems was suggested. Further studies were conducted by the same research group to isolate the bacterium in the distribution systems and to investigate HAAs' biodegradation pathways. Removal of HAAs by granular activated carbon (GAC) filtration was reported. The authors suggested that the reduction was due to biological degradation of HAAs.

Removing trihalomethanes (THMs) using GAC adsorption requires very frequent GAC regeneration and is considered impractical in drinking water treatment. GAC has been believed to be not effective for HAA removal because HAAs are hydrophilic and ionized organics in drinking water. A recent study suggested, however, that carbon adsorption could be an effective process for removing two major HAAs: Cl_2AA and Cl_3AA. The reported Freundlich values for Cl_2AA and Cl_3AA, 1,630 and 11,700 $\mu g/g(L/\mu g)^{1/n}$, respectively, are much higher than that for chloroform ($CHCl_3$) and bromodichloromethane, 93 and 241 $\mu g/g(L/\mu g)^{1/n}$, respectively. A similar study gave q_e (solid concentration at equilibrium) values of 1,600 µg/g for Cl_2AA and 3,900 µg/g for Cl_3AA at a C_e (solution concentration at equilibrium) value of 0.08 µM. However, information on HAA removal using GAC adsorption in batch, pilot, or full-scale studies is very limited.

Biologically active carbon (BAC) or biologically enhanced activated carbon could also effectively remove HAAs in drinking water through biodegradation and carbon adsorption. In a recent study by Zhou and Xie (2002), HAAs underwent significant degradation in the presence of suspended bacteria in dechlorinated drinking water. In a column study, Xie and Zhou (2002) reported that BAC columns effectively removed mono- and dihaloacetic acids spiked at 50 µg/L each in their influents. In another column study with virgin GAC, a nearly complete removal of HAAs was obtained in the first 30 hours of the column run. Between 30 hours and 67 days, the effluent concentrations for HAAs ranged from 10 to 110 µg/L (60%–95% removal) with an influent level of approximately 275 µg/L. After 68 days, a nearly complete removal of HAAs was achieved again, indicating that the removal at the beginning of the run might be due to carbon adsorption. After GAC adsorption capacity was exhausted and before bioactivity was developed, HAAs were not removed. After bioactivity was fully developed, a complete removal was achieved because of HAA biological degradation.

RESEARCH OBJECTIVES

The objectives of the proposed research were

1. To study the mechanisms of HAA removal by GAC filtration
2. To evaluate the effects of water temperature and empty-bed contact time (EBCT) on HAA removal in BAC columns
3. To explore techniques to accelerate bioactivity development in GAC filters.

APPROACHES

The research objectives were accomplished by the following five specific tasks:

1. The first task evaluated the adsorption capacity of GAC for HAAs and THMs through bench-scale tests. Bottle-point adsorption tests were conducted with reagent water and field filter influent samples for six HAAs (HAA6) and THMs. The GAC bed lives of a full-scale GAC filter for HAAs and THMs were estimated using their isotherm data.
2. The goal of the second task was to study the mechanisms of HAA degradation in GAC filters. Various tests were conducted to investigate the effects of temperature stress, disinfectant shock, and dissolved oxygen (DO) on HAA removal in BAC filters. The effects of DO on HAA removal were also conducted in batch tests. The release of inorganic halide during HAA degradation was investigated as well.
3. During the third task, a column study was performed to evaluate effects of EBCT and temperature on HAA removal in BAC filters. The study was conducted at four different temperatures (4°, 10°, 20°, and 30°C) and four different EBCTs (5, 10, 15, and 20 minutes). Kinetic models for HAA removal at various temperatures were explored.
4. The goal of the fourth task was to investigate HAA removal in full-scale GAC filters. HAA levels in influents and effluents of four pairs of full-scale GAC filters were monitored. Each pair of GAC filters had different GAC online ages. The HAA levels in six distribution systems—three with GAC filtration and three without GAC filtration—were also monitored to investigate the effects of GAC filtration on HAA levels in the distribution systems.
5. Finally, an exploratory study was performed to evaluate various techniques to accelerate bioactivity development in GAC filters. The study investigated effects of seeding with acclimated GAC and sand in new GAC filters. The effects of chlorine residual quenching and nitrate addition in the filter influent were also investigated.

CONCLUSIONS

The adsorption isotherm tests were conducted for six HAAs and four THMs. Their isotherm constants were calculated using the Freundlich equation. HAA species with a higher halogen number had a higher GAC adsorption capacity (K). For both HAAs and THMs, their brominated species had higher GAC adsorption capacities (K) than their chlorinated analogues.

The isotherm constants indicate that the GAC adsorption capacities for THMs were higher than for ClAA, BrAA, Cl_2AA, bromochloroacetic acid (BrClAA), and Br_2AA. The full-scale GAC filter study also indicated that the GAC bed life for $CHCl_3$ is longer than that for ClAA,

BrAA, Cl_2AA, BrClAA, and Br_2AA. This is contrary to what was reported in other studies and may be due to HAA biodegradation during adsorption isotherm tests. Biodegradation of HAAs during isotherm tests is likely to result in an overestimation of their adsorption capacity. To improve the accuracy of GAC bed life estimation, the isotherm tests should be conducted in the presence of minimal biological activity and with the actual filter influent.

The adsorption isotherm tests and the full-scale filter study indicate that the physical adsorption only plays a minor role in DBP removal in GAC filters.

The biofilm activity inhibition factors, including temperature stress (4°C and 50°C) and chloramine shock, reduced the HAA removal rates in BAC columns significantly. This suggests that biodegradation is the major mechanism of HAA removal in BAC columns. No effect of anoxic environment on HAA removal was observed in column tests. Results from the batch test showed that HAAs, with the exception of Cl_3AA, were degraded under both aerobic and anoxic conditions. For HAA degradation, however, a lag phase was observed in anoxic conditions. Measurements of chloride released from HAAs incubated with BAC for 80 hours indicated a complete dehalogenation for ClAA, 60% dehalogenation for Cl_2AA, and 15% dehalogenation for Cl_3AA. These results suggest that a large percentage of HAAs were removed through biological dehalogenation during BAC filtration.

Both EBCT and water temperature significantly affected the HAA removal in BAC columns. Increasing the water temperature and EBCT enhances the removal of HAAs. At a temperature of 20°C or higher, increasing EBCT resulted in small increases in HAA removal. Longer EBCTs could enhance HAA removal at lower water temperatures (<10°C). In order to achieve a removal efficiency of 50% or higher for HAA6, the results from this study suggest a 10-minute EBCT for 4°C waters and a 5-minute EBCT for waters at 10°C or higher.

HAA removal in BAC columns had different kinetic models at different temperatures. At 4°C, HAA removal tended to follow a zero-order reaction model. This suggests that HAA removal was limited by carbon adsorption capacity or biological activities on its surface. At 10°C, the mono- and dihaloacetic acid removal tended to follow a first-order reaction model. This suggests that biological activities were fully developed at 10°C and HAA removal was limited by HAA or its concentrations. For Cl_3AA removal, a first-order reaction model was obtained at 20°C but not at 10°C.

GAC filters effectively reduced HAA concentrations in its effluents. The average HAA removal efficiencies for full-scale GAC filters studied ranged between 60% and 85%. There were no significant differences in HAA removal between newer and older GAC filters (6 months to 6½ years older), especially after a few months' operation of a new GAC filter. EBCT affected the HAA removal, especially for Cl_3AA. GAC filters with longer EBCTs could improve HAA removal. The temperature of the influent water could significantly affect HAA removal, especially for Cl_3AA. Less Cl_3AA was removed during BAC filtration in the winter months than during the summertime. The difference in HAA removal was more pronounced in GAC filters with short EBCTs (<10 minutes).

GAC filtration could significantly reduce HAA5 levels in the distribution system. In three GAC water systems, GAC filtration significantly increased the ratios of THMs to HAA5. For two non-GAC systems, HAA5 levels in five of the six samples from distribution systems were higher than the MCL (60 μg/L). In two GAC systems, however, GAC filtration reduced the HAA5 levels below the MCL (60 μg/L) in all six samples, even though THM levels were higher than the MCL (>80 μg/L) in four of the six samples. Even though HAAs and other DBPs are continuously formed in the distribution system, GAC filtration would be a viable technology for HAA5 compliance.

Biological seeding of GAC filter beds, with 5% acclimated GAC, could enhance HAA removal at the early stage of a GAC filter operation. Results show that seeding with 5% acclimated GAC could also reduce the time required for full biological activity development. However, there was no observable impact on Cl_3AA removal as a result of acclimated GAC seeding. The use of acclimated sand as seeding material did not improve the bioactivity development in a GAC filter.

Backwash with chlorinated water at two backwash rates (35.4 and 77.0 m/hr [14.5 and 31.5 gpm/ft^2]) did not significantly affect HAA removal in BAC filters. Chlorine residual quenching of the influent water with sodium thiosulfite resulted in an early HAA breakthrough and delayed the biological activity development. The addition of nitrate to the influent water resulted in no noticeable effect on the biological activity development and biological degradation of HAAs except for an early HAA breakthrough. In both cases, the breakthrough may be due to the competition between HAAs and thiosulfate or nitrate for the GAC adsorption site. Lower HAA concentrations on the GAC surface, caused by thiosulfate competition, may delay the biological activity development on the GAC. The mechanisms of the thiosulfate and nitrate effects should be investigated in future studies.

RECOMMENDATIONS

GAC filtration is an effective process for removing HAAs formed during prechlorination or intermediate chlorination. It could also be used by consecutive systems for HAA control in their distribution systems. GAC filtration can provide a long-term (several years) removal for HAAs through biological degradation and a short-term (several months) removal for THMs and Cl_3AA through physical adsorption.

Replacing the GAC should be staggered and preferably done in early summer, which would better utilize the adsorption capacity of new GAC filters to control higher levels of THMs and Cl_3AA in summer months. After the summer, the saturated and biologically active GAC filters could provide a better biological degradation for HAAs. In a new GAC filter, seeding with acclimated GAC will expedite the biological activity development and could significantly benefit HAA removal.

Longer EBCTs could be used to enhance HAA removal at low water temperatures. In order to achieve a removal efficiency of 50% or higher for HAAs, the authors suggest that a 10-minute EBCT be used for 4°C waters (in the coldest month) and a 5-minute EBCT for waters 10°C or higher.

Biodegradation and organic matrix competition during the isotherm test for HAAs can greatly affect their isotherm constants. The isotherm tests for HAAs should be conducted in the presence of minimal biological activity with the actual filter influent. For HAA removal by physical adsorption, adsorption isotherm data obtained under these conditions provide a better GAC bed life estimation for full-scale GAC filters or contactors.

CHAPTER 1
INTRODUCTION

OBJECTIVES

Haloacetic acids (HAAs) are a group of organic disinfection by-products (DBPs) regulated under the Stage 1 Disinfectants/Disinfection By-Products (D/DBP) Rule (USEPA 1998). The maximum contaminant level (MCL) for the sum of five haloacetic acids (HAA5) under the promulgated rule is 60 µg/L. These five HAAs are monochloroacetic acid (ClAA), monobromoacetic acid (BrAA), dichloroacetic acid (Cl_2AA), dibromoacetic acid (Br_2AA), and trichloroacetic acid (Cl_3AA). Enhanced coagulation/softening for precursor removal and alternative disinfectants are common means for controlling HAAs and other DBPs in finished water (Krasner and Amy 1995, Singer 1994).

It is well documented that biological filtration effectively removes ozonation by-products and some natural organic matter (NOM) (Krasner, Sclimenti, and Coffey 1992; Miltner and Summers 1992; Reckhow et al. 1992; Xie and Reckhow 1992; Hozalski, Bouwer, and Goel 1999; Huck 1999; Xie 2004). However, limited information is available on the use of biofiltration for HAA removal. In 1993, Singer et al. (1993) reported a rapid elimination of HAAs in treated water during aquifer storage. A biological degradation mechanism was suggested. In 1994, Williams, Williams, and Rindfleisch (1994) reported exceptionally low levels of HAAs at the maximum residence time locations in three distribution systems in Virginia. One explanation reported involves the possible bacterial metabolism or degradation of HAAs. Further studies were conducted by the same research group to isolate the bacterium in the distribution systems and to investigate biodegradation pathways of HAAs (Williams, Rindfleisch, and Williams 1995; Williams, Williams, and Gordon 1997; Williams, Williams, and Yuan 1998). Removal of HAAs by granular activated carbon (GAC) filtration was reported, presumably by biodegradation processes with the filter bed (Singer et al. 1999).

GAC adsorption for removing trihalomethanes (THMs) requires very frequent regeneration and is considered impractical in drinking water treatment (Graese, Snoeyink, and Lee 1987). It had been commonly assumed that GAC is not effective for HAA removal because HAAs are hydrophilic and ionized organics in drinking water. However, a recent study suggested that carbon adsorption could be an effective process for removing two major HAAs: Cl_2AA and Cl_3AA (Speth and Miltner 1998). The reported Freundlich values for Cl_2AA and Cl_3AA, 1,630 and 11,700 $\mu g/g(L/\mu g)^{1/n}$, respectively, are much higher than those for chloroform ($CHCl_3$) and bromodichloromethane ($CHBrCl_2$), 93 and 241 $\mu g/g(L/\mu g)^{1/n}$, respectively (Speth and Miltner 1998). A similar study gave q_e (solid concentration at equilibrium) values of 1,600 µg/g for Cl_2AA and 3,900 µg/g for Cl_3AA at a C_e (solution concentration at equilibrium) value of 0.08 µM (Zhang et al. 1998). However, very limited information is available on HAA removal using GAC adsorption in batch, pilot, or full-scale studies.

In a recent study by Zhou and Xie (2002), HAAs underwent significant degradation in the presence of suspended bacteria in dechlorinated drinking water. Biologically active carbon (BAC) or biologically enhanced activated carbon also effectively removed HAAs in dechlorinated drinking water through biodegradation and carbon adsorption. In a column study, Xie and Zhou (2002) reported that BAC columns effectively removed mono- and dihaloacetic acids spiked at 50 µg/L each in their influents. In another new GAC column study, in the first 30 hours of the column run, a nearly complete removal of HAAs (less than 5 μg/L in effluents) was obtained (Xie and Zhou 2002). Between 30 hours and 67 days, the HAA concentrations in the effluent ranged from 10 to 110 μg/L (60% to 95% removal) with an influent level of approximately 275 μg/L. After 68 days, a nearly complete removal of HAAs was achieved again. This indicated that the removal at the beginning of the run might be due to carbon adsorption. After GAC adsorption was exhausted and before bioactivity was developed, HAAs were not removed. After bioactivity was fully developed, a complete removal was achieved because of HAA biological degradation.

RESEARCH OBJECTIVES

The objectives of the proposed research were

1. To study the mechanisms of HAA removal by GAC filtration
2. To evaluate the effects of water temperature and empty-bed contact time (EBCT) on HAA removal in BAC columns
3. To explore techniques to accelerate bioactivity development in GAC filters.

APPROACHES

The research objectives were accomplished by the following five specific tasks, each of which is discussed in chapters 2 through 6:

1. The first task evaluated the adsorption capacity of GAC for HAAs and THMs through bench-scale tests. Bottle-point adsorption tests were conducted with reagent water and field filter influent samples for six HAAs (HAA6) and THMs. The GAC bed lives of a full-scale GAC filter for HAAs and THMs were estimated using their isotherm data.
2. The goal of the second task was to study the mechanisms of HAA degradation in GAC filters. Various tests were conducted to investigate the effects of temperature stress, disinfectant shock, and dissolved oxygen (DO) on HAA removal in BAC filters. The effects of DO on HAA removal were also conducted in batch tests. The release of inorganic halide during HAA degradation was investigated as well.
3. During the third task, a column study was performed to evaluate effects of EBCT and temperature on HAA removal in BAC filters. The study was conducted at four different temperatures (4°, 10°, 20°, and 30°C) and four different EBCTs (5, 10, 15, and 20 minutes). Kinetic models for HAA removal at various temperatures were explored.
4. The goal of the fourth task was to investigate HAA removal in full-scale GAC filters. HAA levels in influents and effluents of four pairs of full-scale GAC filters were monitored. Each pair of GAC filters had different GAC online ages. The HAA levels in six distribution systems—three with GAC filtration and three without GAC filtration—were also monitored to investigate the effects of GAC filtration on HAA levels in the distribution systems.
5. Finally, an exploratory study was performed to evaluate various techniques to accelerate bioactivity development in GAC filters. The study investigated effects of seeding with acclimated GAC and sand in new GAC filters. The effects of chlorine residual quenching and nitrate addition in the filter influent were also investigated.

ORGANIZATION OF REPORT

This report consists of seven chapters, including this introductory chapter, chapter 1. Chapter 2 examines the effects of adsorption isotherm test conditions, especially biological activities and organic matrix, on the GAC bed life estimation. Chapter 3 presents a preliminary study on the mechanisms for HAA biodegradation. Chapter 4 describes the effects of EBCT and water temperature on HAA removal in BAC columns. Chapter 5 presents the study on HAA removal in full-scale GAC filters. Chapter 6 presents a preliminary study on biofilm development in GAC filters. Chapter 7 summarizes the conclusions from all five studies or chapters and expresses the implications as well as what future studies are needed.

CHAPTER 2
THE EFFECTS OF ADSORPTION ISOTHERM TESTING CONDITIONS ON GAC BED LIFE ESTIMATION

INTRODUCTION

Speth and Miltner (1990, 1998) reported various GAC adsorption isotherm studies for many organic compounds, including two HAAs and the sum of four trihalomethanes (THM4), in organic-free water. Freundlich constants (*K*) for THMs were reported to be 92.5, 241, 585, and 929 $\mu g/g(L/\mu g)^{1/n}$ for $CHCl_3$, dichlorobromomethane ($CHCl_2Br$), chlorodibromomethane ($CHBr_2Cl$), and bromoform ($CHBr_3$), respectively (Speth and Miltner 1990). The constants for Cl_2AA and Cl_3AA were 1,630 and 11,700 $\mu g/g(L/\mu g)^{1/n}$, respectively (Speth and Miltner 1998). According to these constants, GAC adsorption could be an efficient treatment process for HAA control. The calculated GAC bed lives from those constants for HAAs (e.g., approximately 2 years for Cl_2AA, 3 years for Cl_3AA) are significantly longer than those for THM species (e.g., 2–3 months for $CHCl_3$). However, much shorter GAC bed lives (in the range of months) for HAAs have been reported. For Cl_2AA, the GAC bed life of a laboratory GAC column was approximately 1 month (Xie and Zhou 2002). Another pilot column study conducted in a water treatment plant also found shorter GAC bed lives for HAAs (13 days for Cl_2AA and 33 days for Cl_3AA; Xie, Wu, and Tung 2003).

Adsorption test results were affected by the content of organic matter in the aqueous phase. The presence of NOM usually reduces the GAC adsorption capacity for the targeting adsorbate. A multisolute system was shown to compete for the carbon adsorption site or to cause desorption of the weaker adsorbed compound (Baker 1964). In adsorption tests, it is important to select a proper aqueous phase that can represent the actual condition of GAC contactor operation for GAC bed life estimation. However, organic-free water is commonly used for preparing adsorbate solution when conducting laboratory adsorption tests.

The purpose of this task is to develop a proper experimental procedure to estimate GAC bed lives for HAAs during water treatment. Bottle-point adsorption isotherm tests were conducted to obtain GAC adsorption capacity. Since HAAs are subject to biodegradation, microbial activities can reduce the liquid-phase concentration and result in an overestimation of the adsorption capacity. With the biological activities controlled in this study, the isotherm results were shown to provide more accurate GAC bed life estimations for HAA removal by physical adsorption. The performance of a full-scale GAC filter in removing HAAs and THMs was also monitored to verify the GAC bed life estimation using the adsorption isotherm data.

MATERIALS AND METHODS

Chemicals

Pure HAAs and $CHCl_3$ were used to prepare solutions for bottle-point adsorption tests. $CHCl_3$ and HAA6—ClAA, Cl_2AA, Cl_3AA, BrAA, Br_2AA, and bromochloroacetic acid (BrClAA)—and 1,2-dibromopropane were obtained from Aldrich Chemical Co., Milwaukee, Wis. The purities of these chemicals were certified at 97% or higher. For THM and HAA analysis, mixed THM4 (2,000 µg/mL each in methanol) and HAA6 (2,000 µg/mL each in methyl-*tert*-butyl-ether [MTBE]) calibration standards were purchased from Supelco, Bellefonte, Pa.

Adsorption Isotherms

Each HAA solution was prepared in both organic-free water and GAC filter influent. This filter influent was collected from a local water treatment plant to simulate the organic content in the field operation. The average total organic carbon concentrations for this filter were 1.56 and 0.92 mg/L for the field filter influent and effluent, respectively. All solutions were sterilized using a 0.22-µm filter. Regenerated GAC (Filtrasorb F-400; Calgon Carbon Corp., Pittsburgh, Pa.), as used in the field GAC filter, was washed with organic-free water and dried at 105°C before use. Sterilized bottles filled with HAA solutions and GAC were placed horizontally on a reciprocal shaker (100 rpm) and equilibrated for 24 hours. Water samples were then collected after being filtered through a 0.45-µm glass fiber syringe filter. For THM samples, similar protocols were used, except these bottles were sealed with polytetrafluoroethylene (PTFE)/butyl septa without any headspace.

The isotherm constants were calculated by the Freundlich equation:

$$q_e = KC_e^{1/n}$$

where q_e = solid concentration at equilibrium (µg/g)
K = Freundlich constants [µg/g(L/µg)$^{1/n}$]
C_e = liquid-phase concentration at equilibrium (µg/L)
$1/n$ = constant, dimensionless

The C_e was the actual equilibrium concentration measured from each water sample, and the q_e was calculated using the following equation:

$$q_e = \left(\frac{V}{M}\right)(C_0 - C_e)$$

where V = reactor volume (L)
M = mass of carbon (g)
C_0 = initial concentration (µg/L)

Field Filter Study

The field study was conducted on a newly replaced GAC filter in the local water treatment plant. This filter is a GAC–sand dual-media filter with an average flow of 3.4 m/hr (1.4 gpm/ft^2). The EBCT through the GAC media is no less than 10 minutes at maximum flow (6.6 m/hr [2.7 gpm/ft^2]). After 36 to 72 hours of filtration, the filter was backwashed using finished water. Filter influent and effluent samples were collected regularly after the GAC replacement. Samples were dechlorinated with ammonium chloride and transported to Penn State Harrisburg's (Middletown, Pa.) environmental engineering laboratories for analysis.

GAC Bed Life Estimation

GAC bed lives for GAC columns and the field filter were estimated as follows:

$$\text{bed life} = \left[\frac{\text{calculated GAC capacity}}{\text{substrate loading rate}}\right]$$

where calculated GAC capacity (μg) = q_e (μg/g) × carbon mass (g)
substrate loading rate (μg/hr) = influent flow rate (L/hr) × concentration (μg/L)

For the field filter GAC bed life calculation, the substrate loading rates were obtained by integrating the influent concentration versus time from actual sample analyses.

DBP Analysis

HAA concentrations for all the samples were analyzed using a modified USEPA Method 552.2 (USEPA 1995a; Xie, Reckhow, and Springborg 1998; Xie et al. 2002). A Hewlett Packard (Wilmington, Del.) 6890 gas chromatography–electron capture detector (GC–ECD) and a DB-1701 capillary column (30 m × 0.25 mm ID, 0.25 μm film thickness) from J&W Scientific (Folsom, Calif.) were used for HAA analysis. THM concentrations were analyzed using a modified USEPA Method 551.1 (USEPA 1995b). An Agilent (Wilmington, Del.) 6890N GC–ECD and an Agilent HP-1 capillary column (30 m × 0.32 mm ID, 1.0 μm film thickness) were used for THM analysis.

RESULTS AND DISCUSSION

Adsorption Isotherms

Adsorption isotherms constructed for GAC filter influent are presented in Figure 2.1. Freundlich constants (K and $1/n$) for filter influent (FI) and organic-free water (OFW) were calculated and listed in Table 2.1. K values for isotherms obtained for the filter influent (K_{FI}) exhibited the same trend as that observed for isotherms of organic-free water (K_{OFW}). HAA species with a higher halogen number had a larger adsorption capacity (K) for GAC. This trend is similar to that found in the literature (LT) by comparing the Cl_2AA and Cl_3AA data (K_{LT}, Table 2.1, column c) reported by Speth and Miltner (1998). Brominated HAAs have higher GAC adsorption capacities (K) than their chlorinated analogues.

From the isotherm tests, K_{FI} values were significantly less than K_{OFW} values. This reduced adsorption capacity with filter influent may be due to competition for adsorption sites between the organic matrix and HAAs. Although both K_{OFW} and K_{LT} were obtained from the isotherm tests prepared with organic-free water, values of K_{OFW} were significantly lower than those of K_{LT}. This seems to be the result of HAA biodegradation during the isotherm studies. According to the literature, a 3-week equilibrium time was used (Speth and Miltner 1998), which was much longer than that employed in our isotherm studies (24 hours). Since HAAs are biologically unstable, a long equilibrium time may result in biodegradation of HAAs. An equilibrium test indicated that the equilibrium time for HAA adsorption with virgin GAC under our experimental conditions was about 5 hours, as shown in Figure 2.2. The $1/n_{FI}$ values were considerably larger than $1/n_{OFW}$ and

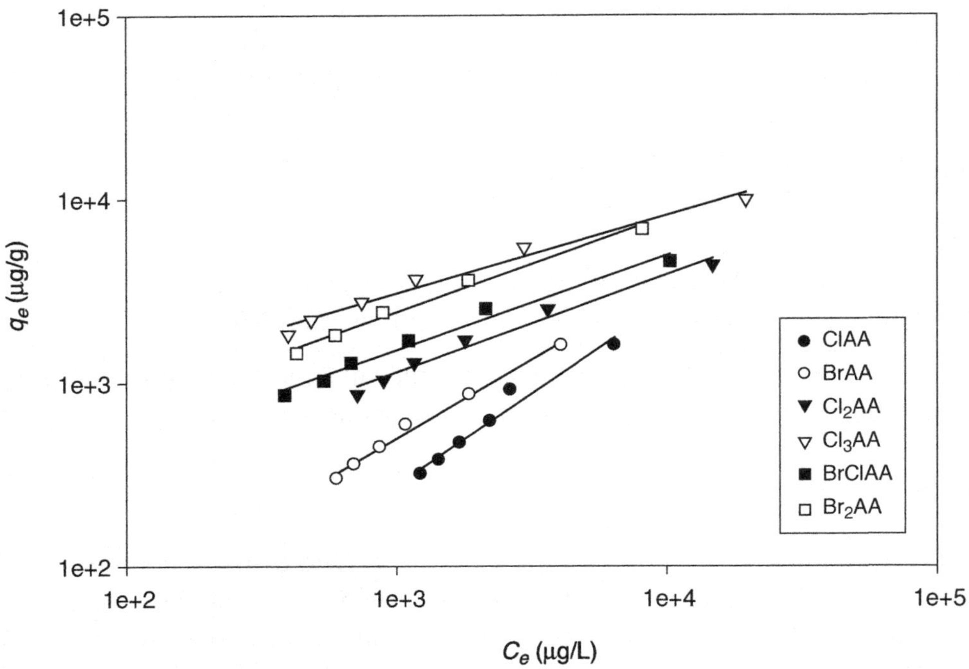

Source: Tung, Unz, and Xie 2003.

Figure 2.1 Bottle-point isotherms for HAAs in field filter influent

Table 2.1
Freundlich constants for HAA6

	a. Organic-free water		b. Filter influent		c. Literature*	
HAA	K_{OFW}†	$1/n_{OFW}$	K_{FI}	$1/n_{FI}$	K_{LT}	$1/n_{LT}$
ClAA	0.43	0.784	0.29	0.993	NA‡	NA
Cl_2AA	208.83	0.300	30.14	0.527	1,630	0.462
Cl_3AA	704.53	0.250	166.53	0.422	11,700	0.216
BrAA	94.93	0.355	1.34	0.857	NA	NA
Br_2AA	504.89	0.288	50.86	0.554	NA	NA
BrClAA	338.14	0.295	44.30	0.511	NA	NA

Source: Tung, Unz, and Xie 2003.
*Speth and Miltner 1998.
†The unit for constant K is $\mu g/g(L/\mu g)^{1/n}$.
‡NA: not available.

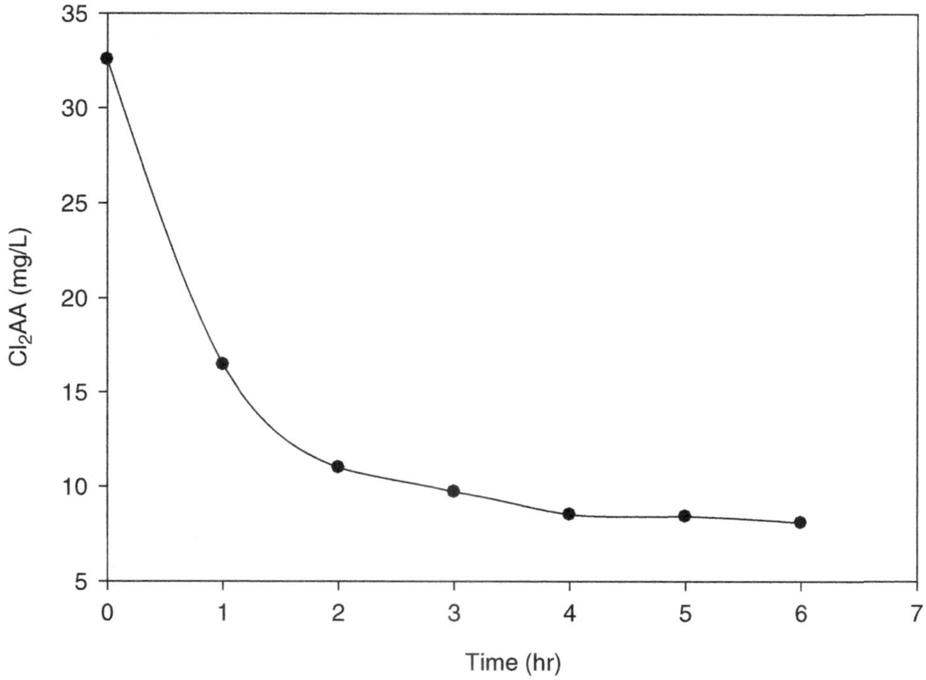

Figure 2.2 Cl_2AA concentration versus time in a single adsorption testing vessel

$1/n_{LT}$ values, possibly indicating that bonding strength between HAAs and GAC was weaker in the field filter influent than in organic-free water. This also indicates that HAAs and the organic matter in the field filter influent competed for sites on GAC.

GAC Bed Life Estimations for HAAs

For HAA removal by physical adsorption, GAC bed lives estimated by adsorption isotherms and observed from the field study and laboratory column study are listed in Table 2.2. All the GAC bed lives were calculated from total adsorption capacity (determined using Freundlich constants and equilibrium concentrations observed in the field filter study) and the HAA loading rates observed in the field study. The GAC bed lives were calculated only for four HAA species—ClAA, Cl_2AA, Cl_3AA, and BrClAA—observed in the filter influent. Only ClAA and Cl_2AA saturation was observed in this field filter. The GAC bed lives observed for ClAA and Cl_2AA (Table 2.2, column e) were slightly longer but close to GAC bed lives estimated by isotherms determined for filter influent (Table 2.2, column a).

The GAC bed lives observed from a previous laboratory GAC column study are shown in Table 2.2, column f (Xie and Zhou 2002). This study was operated under similar conditions as the field filter but with a 20-minute EBCT. The column influent was prepared using finished water spiked with HAA5. By calculating the GAC bed lives using GAC isotherm constants (Table 2.1, column b) and actual HAA loading concentrations, the calculated GAC bed lives (Table 2.2, column d) were close to the GAC bed lives actually observed by Xie and Zhou (Table 2.2, column f).

Table 2.2
Observed and estimated GAC bed lives for HAA6 (in days)

	Estimated by isotherms for field filter			Estimated by isotherms for laboratory column	Observed for field filter and laboratory column	
Isotherm constants source	a. Filter influent	b. Organic-free water	c. Literature*	d. Filter influent	e. Field filter study	f. Lab column study†
ClAA	2	2	NA‡	1	10	3
Cl_2AA	21	592	1,661	22	24	32
Cl_3AA	218	597	9,039	128	>94	>76
BrAA	§	§	NA	5	>94	6
Br_2AA	§	§	NA	103	>94	60
BrClAA	155	963	NA	88	>94	NA

*Speth and Miltner 1998.
†Xie and Zhou 2002.
‡NA: not applicable.
§Species was not detected in the influent.

Both ClAA and Cl_2AA are biodegradable; therefore, suspended or attached microorganisms present in the filter system may degrade those compounds. This reduces the mass loading for GAC adsorption and increases the GAC bed life in the field filter operation. GAC bed lives estimated from isotherms generated for organic-free water (Table 2.2, column b) were longer than those observed for Cl_2AA in the field filter study. This may be due to competition between HAAs and the organics matrix in the field water. GAC bed lives calculated from adsorption isotherms taken from the literature (Table 2.2, column c) are distinctly longer than those obtained for our organic-free water and filter influent. This may reflect possible biodegradation effects during the 3-week isotherm tests reported in the literature. HAA biodegradation might decrease the HAA concentration in the aqueous phase and lead to an overestimate of GAC adsorption capacity. The 24-hour equilibrium time and filter sterilization were used in our filter influent and organic-free water isotherm determination, and these conditions could minimize the biodegradation of HAAs.

Evidence of HAA Biodegradation in the GAC Filter

A typical HAA removal pattern for a new GAC filter is profiled in Figure 2.3 (filter effluent pattern). Low HAA effluent concentrations at the beginning of the filter run indicate GAC adsorption. After the breakthrough point, the effluent concentration increases until the GAC is completely saturated. After a few days of negligible HAA removal, effluent concentration gradually reduces to zero, again corresponding with biofilm development on the GAC surface and associated HAA biodegradation. This trend was observed in a previous GAC column study (Xie and Zhou 2002). Similar removal patterns were also observed for both ClAA and Cl_2AA in the field

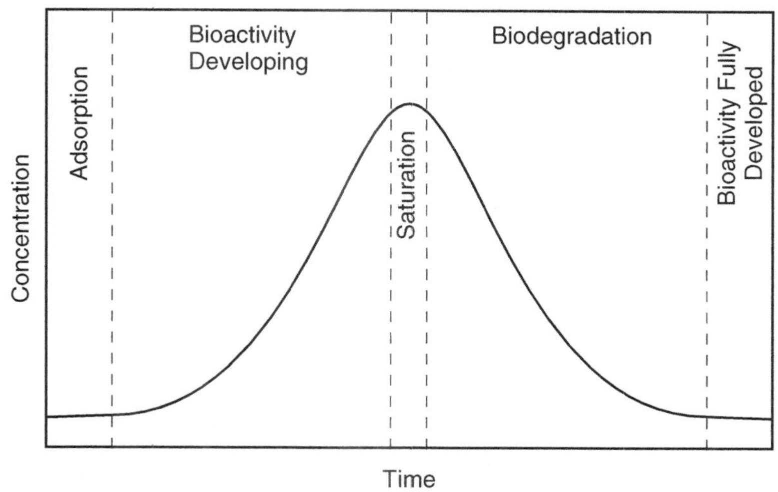

Figure 2.3 Typical HAA removal profile in a new GAC filter

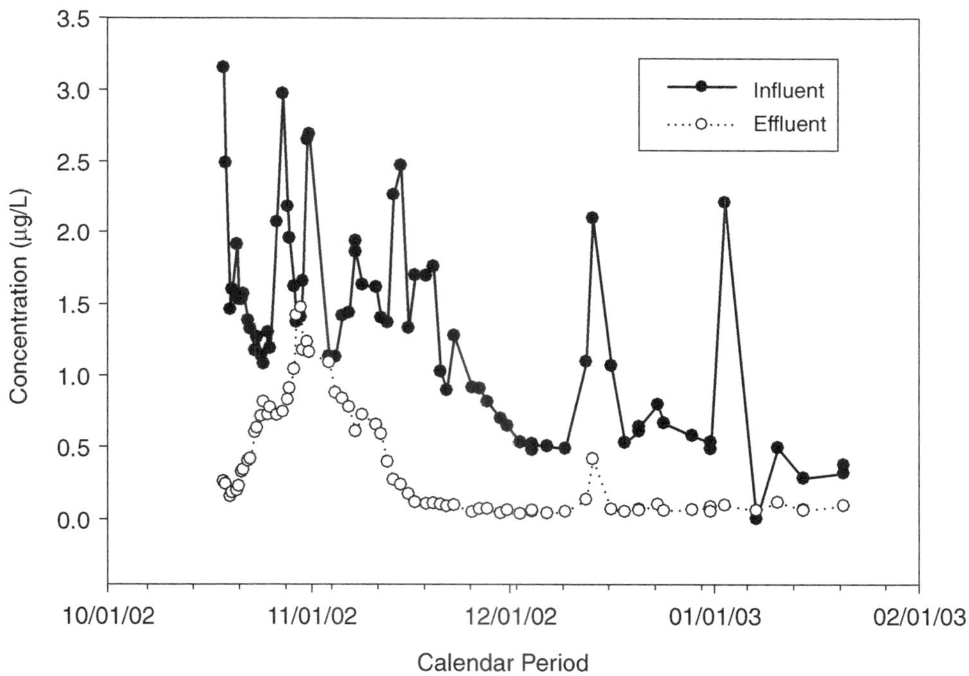

Source: Tung, Unz, and Xie 2003.

Figure 2.4 ClAA concentration profile in field filter study

filter study (Figures 2.4 and 2.5). Because Cl_2AA has a higher adsorption capacity and is slightly less biodegradable than ClAA, its breakthrough and complete removal lagged behind ClAA. The fact that the peak Cl_2AA effluent concentration did not reach the saturation point may be due to biodegradation. Biofilm may already be developed in the system before complete saturation of the GAC with Cl_2AA. In the case of Cl_3AA and other HAAs, no breakthrough was observed in this study. This may be due to their high adsorption capacities or low influent loadings (Figure 2.6).

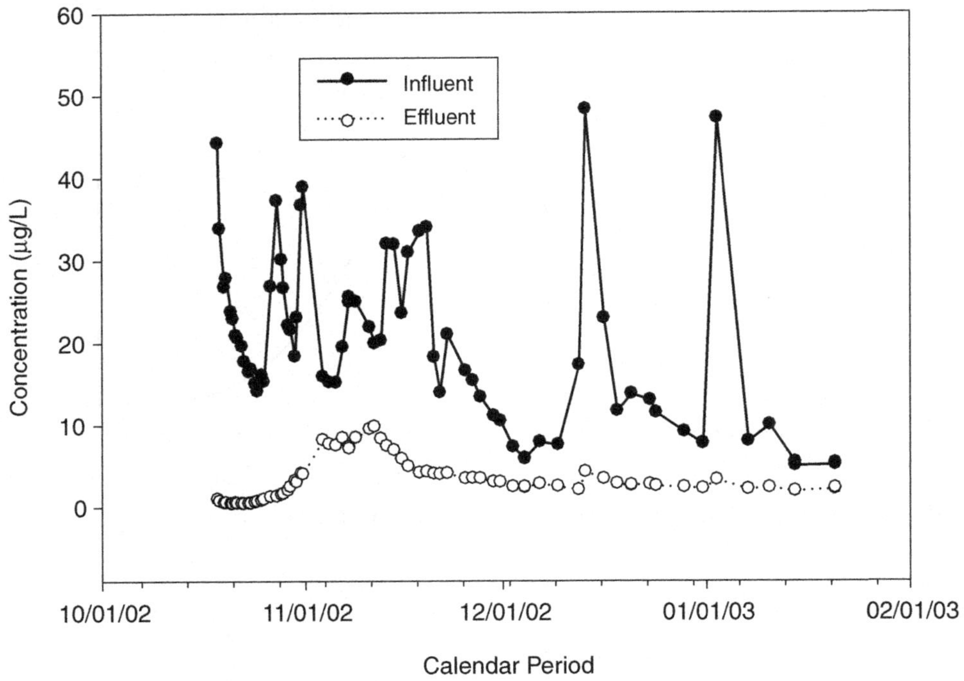

Source: Tung, Unz, and Xie 2003.

Figure 2.5 Cl_2AA concentration profile in field filter study

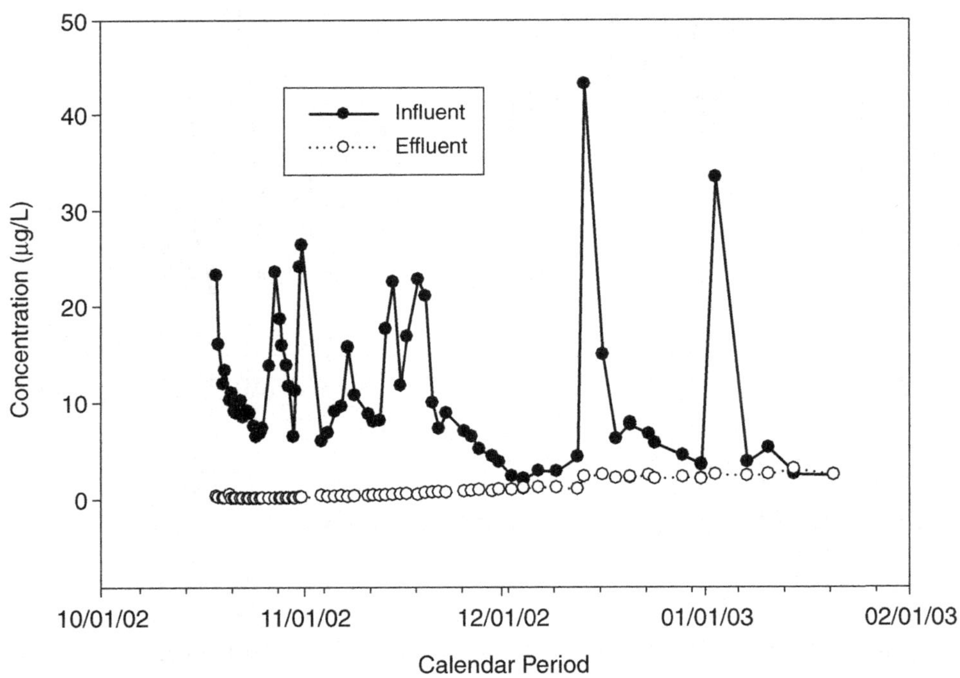

Source: Tung, Unz, and Xie 2003.

Figure 2.6 Cl_3AA concentration profile in field filter study

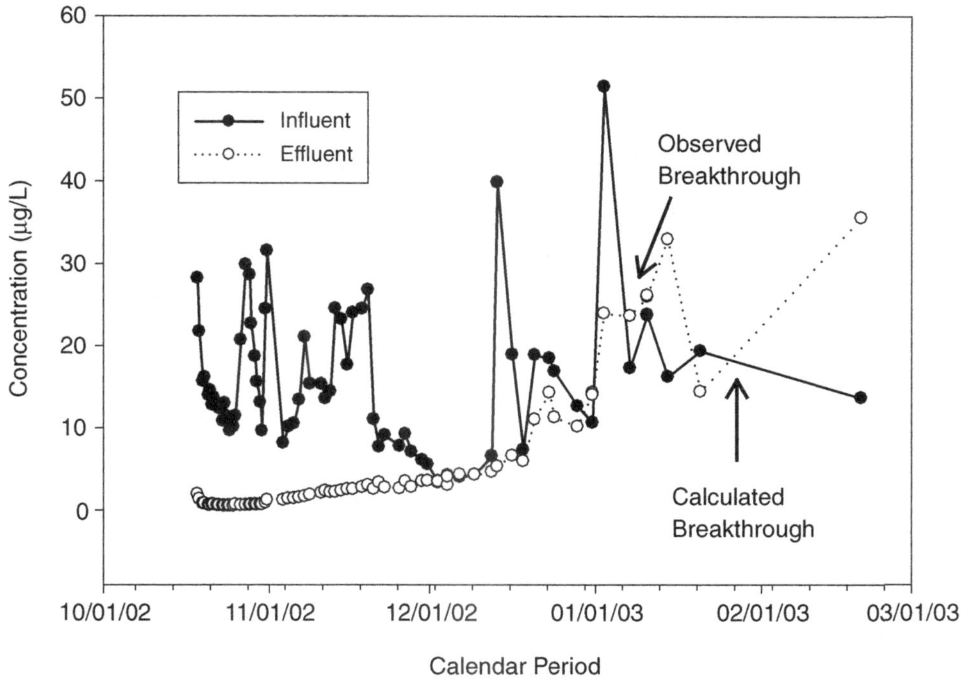

Source: Tung, Unz, and Xie 2003.

Figure 2.7 $CHCl_3$ concentration profile in field filter study

GAC Bed Life Estimation for THMs

In the field operation, only $CHCl_3$ (Figure 2.7) and $CHBrCl_2$ (Figure 2.8) were detected in the filter influent. Effluent $CHCl_3$ concentrations approached the influent concentrations and fluctuated after 88 days of operation. $CHCl_3$ concentrations were higher in the effluent than those in the influent at some points, possibly due to GAC saturation and subsequent desorption of the $CHCl_3$ in varying water quality and operating conditions (Figure 2.7).

The Freundlich constants and GAC bed lives (BL) for THMs are listed in Table 2.3. Similar to HAAs, brominated THMs have higher GAC adsorption capacities (K) than $CHCl_3$. GAC bed lives were calculated using the field filter average influent concentration as C_e for both BL_{LT} and BL_{FI}. GAC bed life for $CHCl_3$, estimated from isotherms for filter influent, was 99 days—close to the actual GAC bed life (around 88 days).

According to published adsorption isotherm data, GAC adsorption capacity for HAAs is higher than for THMs (Speth and Miltner 1990, 1998). For instance, the GAC bed life for Cl_2AA is approximately 9.5 times longer than $CHCl_3$, according to the isotherm data. Applying these values to a water source with similar concentrations of THMs and HAAs, a faster breakthrough for THM species should be observed in GAC contactors. However, under our current observation, the GAC bed life for $CHCl_3$ is longer than for two HAAs (ClAA and Cl_2AA). Another adsorption study was conducted with a commercial point-of-use carbon filter with spiked HAA5 and THM4 as filter influent. The results also show that the carbon bed life for HAA5 was less than for THM4 (Tung et al. 2003).

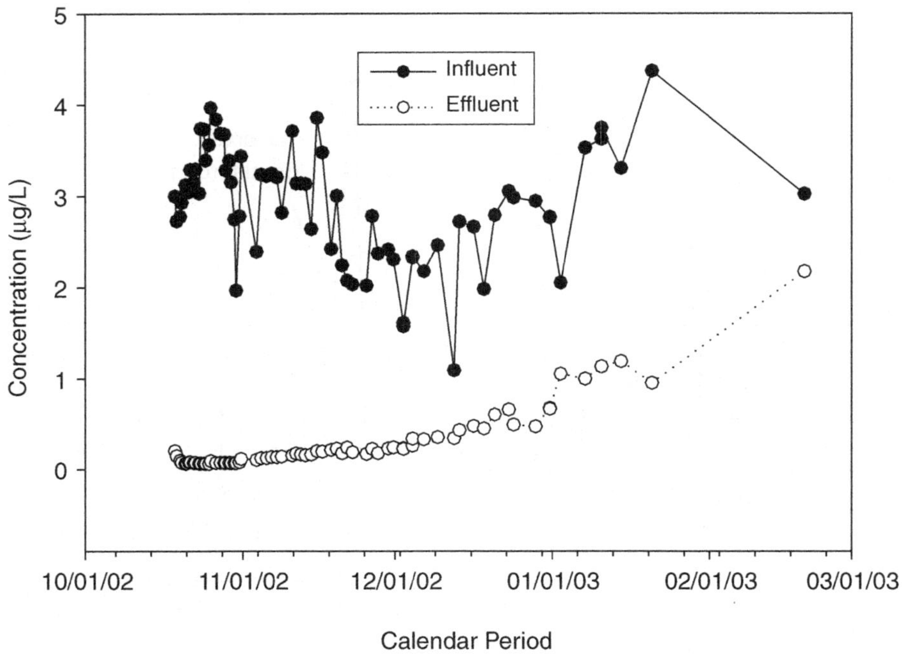

Source: Tung, Unz, and Xie 2003.

Figure 2.8 $CHBrCl_2$ concentration profile in field filter study

Table 2.3
Freundlich constants and estimated GAC bed lives for THMs

	a. Literature*			b. Filter influent			c. Observed field filter	d. Observed literature†
THM	K_{LT}	$1/n_{LT}$	BL_{LT}	K_{FI}	$1/n_{FI}$	BL_{FI}	BL (days)	BL (days)
$CHCl_3$	92.5	0.669	208	68.03	0.544	108	~88	65
$CHBrCl_2$	241	0.655	913	90.95	0.553	313	>124	NA‡
$CHBr_2Cl$	585	0.636	§	116.52	0.625	§	§	NA
$CHBr_3$	929	0.665	§	157.65	0.640	§	§	NA

*Speth and Miltner 1998.
†McGuire and Suffet 1978.
‡NA: not available.
§Species was not detected in the influent.

CONCLUSIONS

The adsorption isotherm tests were conducted for six HAAs and four THMs. Their isotherm constants were calculated using the Freundlich equation. HAA species with a higher halogen number had a higher GAC adsorption capacity (*K*). For both HAAs and THMs, their brominated species had higher GAC adsorption capacities (*K*) than their chlorinated analogues.

The isotherm constants indicated that the GAC adsorption capacities for THMs were higher than for HAAs except Cl_3AA. The full-scale GAC filter study also indicated that the GAC bed life for $CHCl_3$ is longer than that for ClAA, BrAA, Cl_2AA, BrClAA, and Br_2AA. This is contrary to what other studies reported, possibly due to HAA biodegradation during adsorption isotherm tests conducted by others. Such biodegradation will result in an overestimation of their HAAs' adsorption capacity. Conducting the isotherm tests in the presence of minimal biological activity with the actual filter influent will improve the accuracy of GAC bed life estimation for HAA removal by adsorption.

The adsorption isotherm tests and the full-scale filter study indicate that the physical adsorption only plays a minor role in DBP removal in GAC filters.

CHAPTER 3
EVIDENCE OF BIODEGRADATION IN GAC FILTRATION

INTRODUCTION

Several studies revealed that HAAs could undergo biodegradation during water storage or in distribution systems. Singer et al. (1993) examined DBP concentrations during aquifer storage and suggested a biological mechanism for HAA and THM reduction. It was observed that rapid HAA reduction preceded THM degradation, suggesting that HAAs were removed when aerobic conditions were still prevalent in the aquifer, while THM was only removed under anoxic conditions (Singer et al. 1993). Williams, Williams, and Rindfleisch (1994) also suggested that the exceptionally low HAA concentrations at the maximum residence time locations in three distribution systems in Virginia were due to possible biodegradation. HAA concentrations can be higher at the point of entry than at the maximum residence time locations.

Several earlier studies pointed out that certain heterotrophic bacteria are capable of degrading various halogenated aliphatic acids. Jensen (1957) isolated bacteria from soil capable of degrading chloro-aliphatic acids. Weightman, Weightman, and Slater (1992) found *Pseudomonas carboxytroph* and an unidentified methyotroph were capable of liberating chloride from Cl_3AA. Janssen et al. (1985) identified *Xanthobacter autotrophicus* GJ10, which was isolated on 1,2-dichloroethane, as the sole carbon and energy source. This bacterium is also capable of utilizing a number of halogenated short-chain hydrocarbons and halogenated carboxylic acids as the sole carbon source and produces two different dehalogenases (Keuning, Janssen, and Witholt 1985).

GAC is generally used in the drinking water industry to remove taste, odor, and various organic contaminants. The surface area of GAC first provides attachment sites for bacteria where they are protected from high fluid shear forces; the functional groups of the adsorbed organic matter on the surface then chemically interact to bind microorganisms (Carvalho, Vasconcelos, and Bull 2001). Since HAAs can be removed by either adsorption or biodegradation, BAC filtration may be one of the best means to combine both features for HAA reduction. Compared to an anthracite filtration medium, high-rate BAC filters showed a greater reduction efficiency for HAAs (Wobma et al. 2000). Singer et al. (1999) demonstrated that both Cl_2AA and Cl_3AA could be removed by a GAC–sand filter presumably involving a biodegradation process. In the GAC filtration bed, biofilms appear to be active, even when the influent organic concentrations are low. In addition, biofilms are more resistant than suspended growth systems to loading change and toxic materials. Several studies also indicated that BAC was very effective even when the contaminant levels were as low as 1–10 μg/L (Lee, Lu, and Chuang 1994; Shi et al. 1995). A possible mechanism for HAA removal in a GAC filter could be the combination of the GAC adsorption in the initial stage and biodegradation after biofilm activities were fully developed (Xie and Zhou 2002, Zhou and Xie 2002).

The objective of this study is to explore the biological HAA removal characteristics by GAC. The effects of temperature stress, disinfectant, and DO were tested. The results could assist water professionals to better optimize HAA removal using GAC and to comply with the disinfectants and disinfection by-products regulations (USEPA 1998).

MATERIALS AND METHODS

HAA Analysis

Gas Chromatography

HAAs at microgram-per-liter levels were analyzed using modified USEPA Method 552.2 (USEPA 1995a; Xie, Reckhow, and Springborg 1998; Xie et al. 2002). A Hewlett Packard 6890 GC–ECD was used for sample analysis. HAA methyl esters were separated using a DB-1701 capillary column (30 m × 0.25 mm ID, 0.25 μm film thickness).

Capillary Electrophoresis

A Waters Corp. (Milford, Mass.) Quanta 4000 capillary ion analyzer (CIA) was used for milligram-per-liter levels of HAA and halide ion analyses. The CIA configuration includes a negative power supply, a Waters Accu-Sep polyamide-coated fused-silica capillary (60 cm × 75 μm ID), a mercury lamp, and an ultraviolet (UV) absorbance detector. The electrolyte solution consists of 25 mM sodium phosphate and 1.0 mM OFM-OH solution (Waters Corp.). Water samples were filtered through 45-μm nylon filters and analyzed for halide and HAA concentrations by constant current electrophoresis at 45 μA and direct UV detection at 185 nm.

Effects of Thermal Stress

Finished water obtained from a local water treatment plant was spiked with HAA6—ClAA, BrAA, Cl_2AA, BrClAA, Br_2AA, and Cl_3AA—at approximately 100 μg/L each and used as the column influent. Two BAC columns were packed with acclimated GAC collected from the same water treatment plant. One was dedicated for cold shock treatment (column A, 4°C) and the other for heat shock treatment (column B, 50°C). Both columns were equilibrated side by side at 20°C and with an EBCT of 10 minutes for 20 hours. After equilibration, thermal stress was applied to both columns facilitated by water jackets surrounding the columns. Forty-six hours of cold (4°C) and heat (50°C) stress were applied. Both influent and effluent concentrations were monitored intermittently.

Effects of Disinfectant Shock

The column and influent setup were similar to those for temperature shocks. The BAC column was operated at 20°C and equilibrated for 18 hours. After equilibration, the synthetic influent supply was discontinued, and chloramines at 50 mg/L of total chlorine were applied to the column at an EBCT of 2 minutes for 1 hour. During this chloramine shock, the effluent total chlorine concentration was measured to be approximately 5 mg/L, indicating that disinfection was achieved throughout the BAC bed. After the chloramine application, column influent was switched back to the original synthetic influent.

Effects of Dissolved Oxygen Column Test

BAC columns were operated at an EBCT of 10 minutes at room temperature. In the first 19 hours, the influent reservoir was aerated continuously. At the 21st hour, influent was purged by oxygen-free nitrogen, and the DO was maintained at zero. Influent and effluent were sampled intermittently for HAA concentration analysis.

Effects of Dissolved Oxygen Batch Test

Freshly collected BAC was shaken vigorously by hand with dechlorinated finished water and settled for 20 minutes. Supernatant containing washed-off biofilm was used as the microorganism seed. The stock test solution was composed of 2.7 L dechlorinated finished water spiked with HAA6 at 100 μg/L each and mixed with 0.3 L microorganism seed. Aerobic test solutions were dispensed to 125-mL Erlenmeyer flasks or serum bottles (30 mL each). Solutions for anoxic tests were purged with oxygen-free nitrogen and dispensed to 125-mL serum bottles sealed with PTFE/butyl septa (30 mL of test solution in each bottle). Two sets of controls were prepared: one set (blank control) with the finished water and the other set (sterilized control) by autoclaving the seeded solution for 15 minutes. Both controls were spiked with the same concentration of HAA6. Samples were collected for 6 days and analyzed for HAA concentrations. The volatile suspended solids for aerobic and anoxic test solutions and sterilized controls were 147 mg/L, 150 mg/L, and 144 mg/L, respectively.

Release of Halide

The decrease of HAAs and increase of halide were measured in batch incubation. Fifteen grams (wet weight) of BAC were added to 150 mL of distilled water spiked with HAAs. Flasks containing samples were shaken on an orbital shaker, and water samples were taken periodically by filtering through 45-μm nylon filters. HAAs and chloride in these samples were measured using capillary ion electrophoresis.

RESULTS AND DISCUSSION

Effects of Thermal Stress on HAA Removal

In the BAC columns, HAA removal efficiencies at 20°C were close to 100%, except for Cl_3AA. Total influent and effluent HAA6 concentrations from the temperature shock studies are shown in Figure 3.1. HAA removal efficiencies were greatly reduced under both thermal stresses, 4°C and 50°C. Because both temperatures inhibit biological activities, the reduction in HAA removal efficiency suggests that biological activity may play a major role in HAA removal.

Effluent HAA concentrations from both columns were close to zero before thermal stress application (0 to 20th hour). The HAA concentration rose more rapidly in the cold shock column than in the heat shock column. For Cl_2AA, the average removal efficiency was reduced from 98% at 20°C to 25% at 4°C, and then 29% at 50°C, respectively, as shown in Figure 3.2. This suggests that the biofilm involved in HAA degradation may be more susceptible to low temperature than to high temperature.

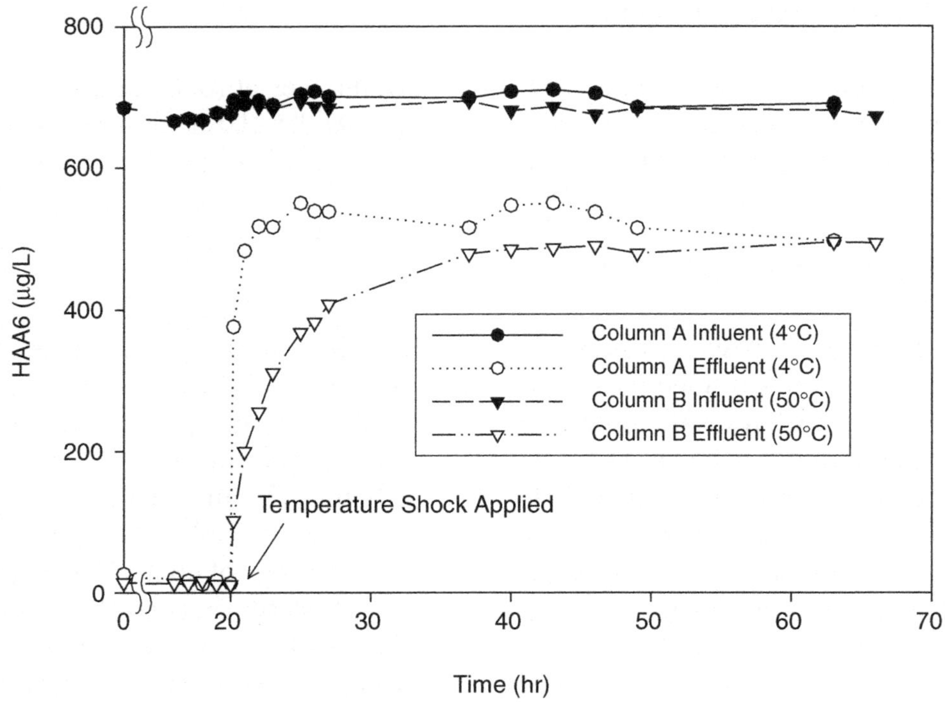

Figure 3.1 Effects of temperature stress on effluent HAA concentrations

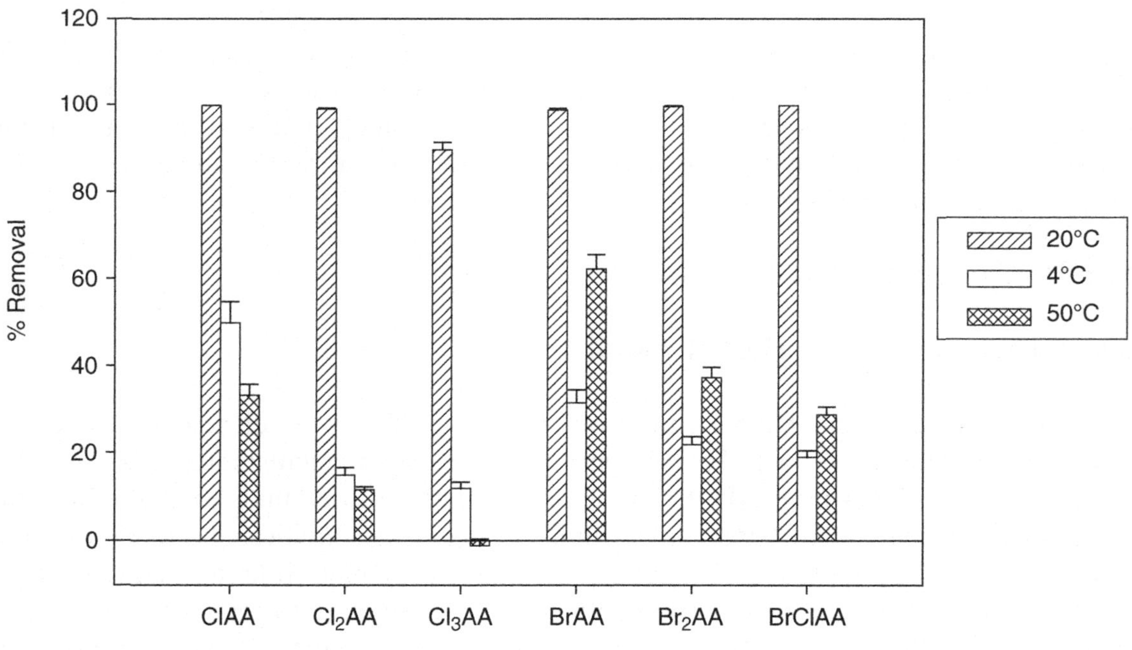

Figure 3.2 Effects of temperature stress on HAA removal

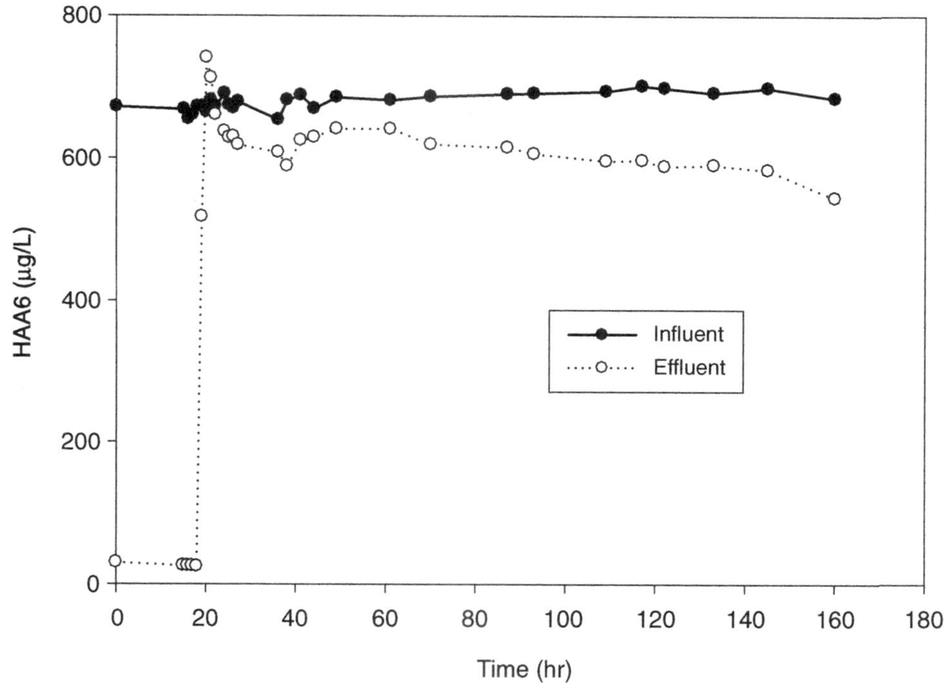

Figure 3.3 Effects of chloramine shock on effluent HAA concentrations

The reduction in removal efficiency obtained under both temperature shocks was significant for most HAA species, especially Cl_3AA. Previous results indicated that Cl_3AA was more resistant to biological degradation than other species were. The greater temperature impact on Cl_3AA removal observed in this study seems to contradict previous studies and needs to be further investigated. A possible reason could be that the bacteria involved in the Cl_3AA degradation are more fragile and susceptible to temperature changes.

Effects of Disinfectant Shock

Figure 3.3 shows the HAA6 concentration before and after the chloramine shock. In the first 18 hours, a nearly complete removal of HAAs was achieved. After the chloramine shock, which is intended to eliminate the biofilm activity, no HAA removal was observed. This suggests that biofilm activity played a major role in HAA removal.

Right after the chloramine shock, the effluent concentration increased to a level higher than the influent concentration. This may be due to the formation of HAAs with chloramines or desorption of HAAs from the biofilm or the GAC. Within hours after the chlorine shock, the effluent concentrations remained high, and no HAA removal was observed. After several days, the effluent concentration decreased. This indicates that the biofilm was recovering from the disinfectant shock and starting to degrade HAAs.

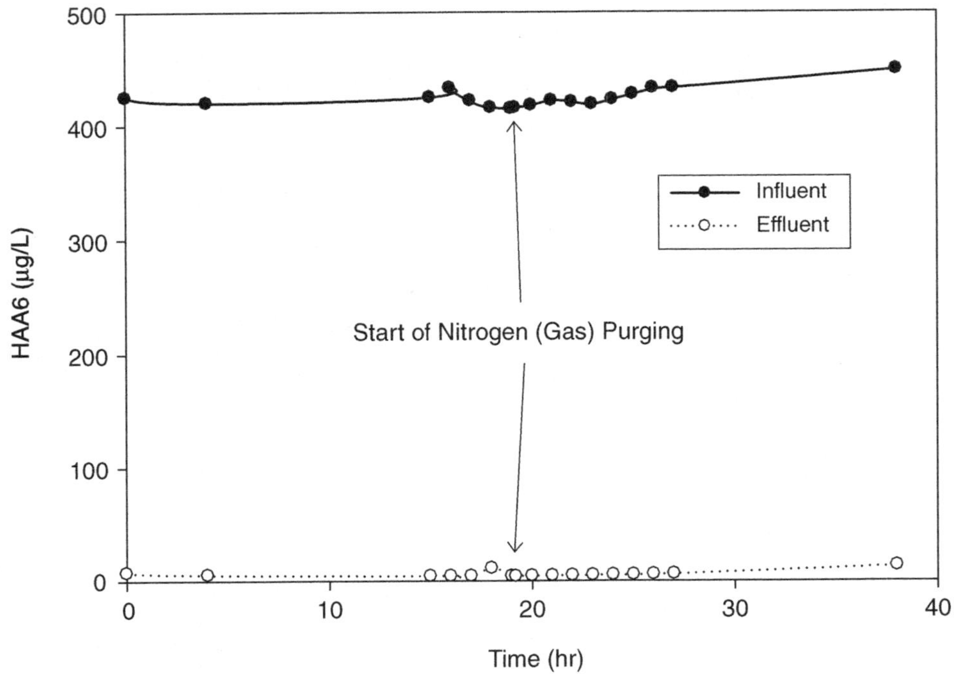

Source: Tung, Unz, and Xie 2003.

Figure 3.4 HAA removal in BAC column

Effects of Dissolved Oxygen Column Test

The purpose of this experiment was to test the BAC column's response to an anoxic environment. No effect on the HAA removal efficiency was observed when oxygen was depleted in the influent (Figure 3.4). This suggests that HAA removal was not affected by the oxygen-limited condition. It is possible that HAA degradation takes place in an anoxic zone of the biofilm or in a certain depth of the GAC filter bed, which is low in DO.

Effects of Dissolved Oxygen Batch Tests

To further explore the reactions and characteristics of HAA biodegradation, batch tests were conducted in both aerobic and anoxic conditions. In the aerobic test, HAAs disappeared rapidly in the first 24 hours (Figures 3.5–3.10). All HAAs were almost completely degraded in both aerobic and anoxic tests, except for Cl_3AA in the anoxic culture. However, HAA biodegradation rates in anoxic conditions were slower than those under aerobic conditions. The removal of Cl_3AA during anoxic incubation was only 45% after 170 hours' incubation. By contrast, in the previous anoxic column test, the BAC column still removed 96% of Cl_3AA after being switched to anoxic conditions. This may also indicate that bacteria species for Cl_3AA degradation are fragile and more susceptible to environmental changes, as described under thermal stress. The apparent lag phase for all HAAs under anoxic conditions may be indicative of the longer time required for bacteria to adapt to the new anoxic conditions. In addition, the biofilm wash-off process could change the microbial ecology and reduce the reduction rate for Cl_3AA and other HAAs.

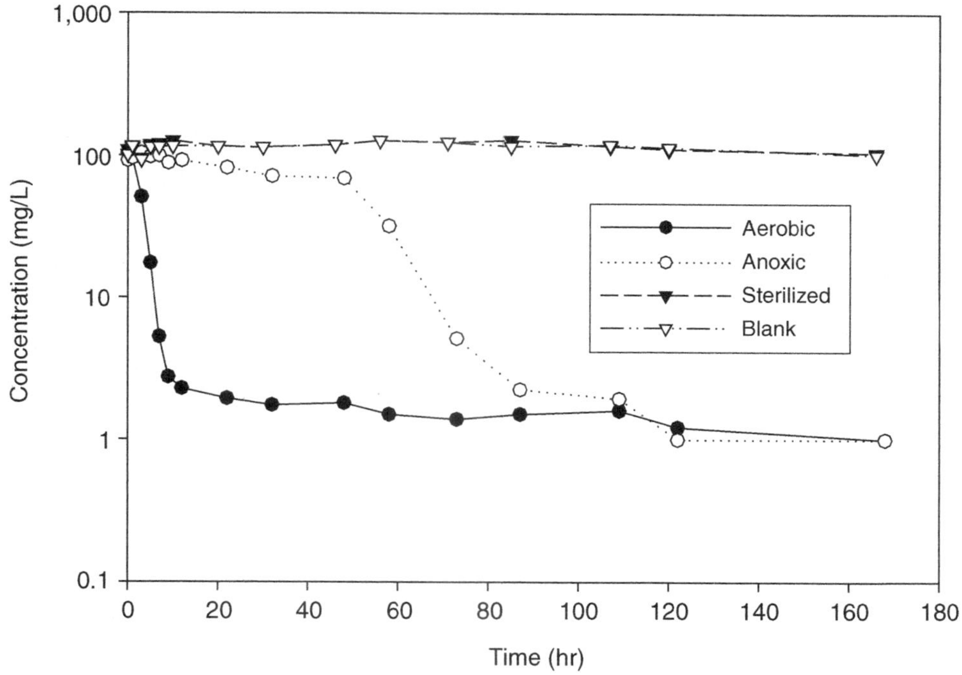

Figure 3.5 ClAA concentration in batch incubations

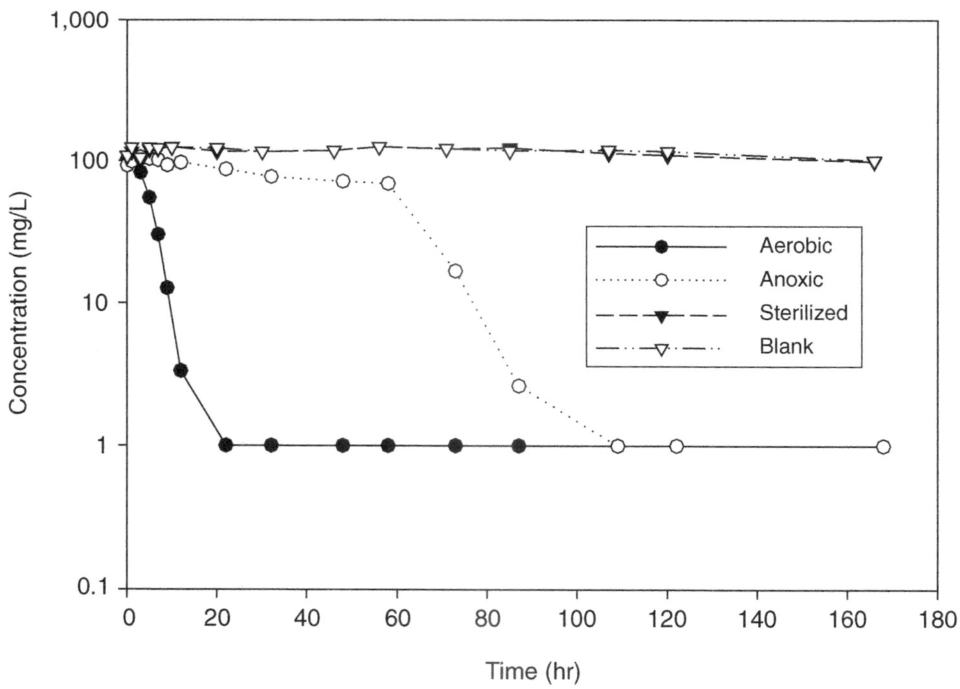

Figure 3.6 BrAA concentration in batch incubations

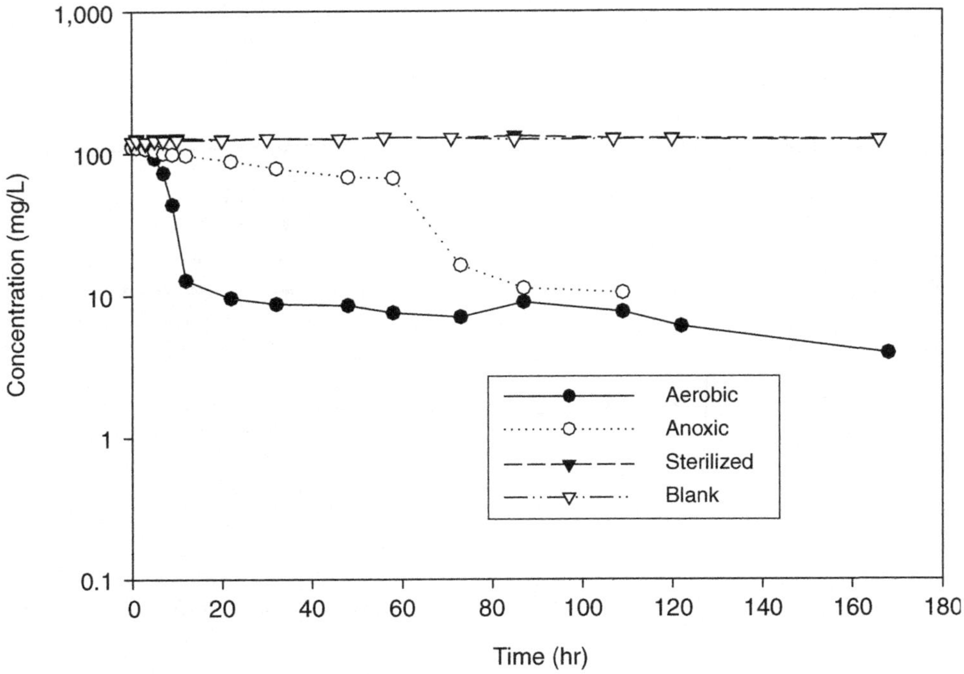

Figure 3.7 Cl_2AA concentration in batch incubations

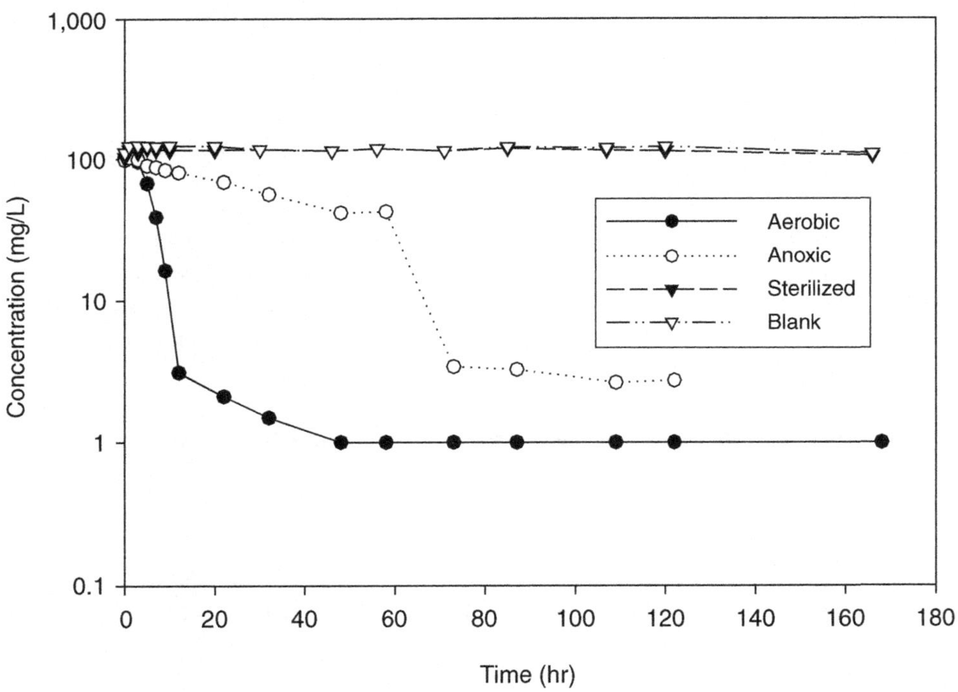

Figure 3.8 Br_2AA concentration in batch incubations

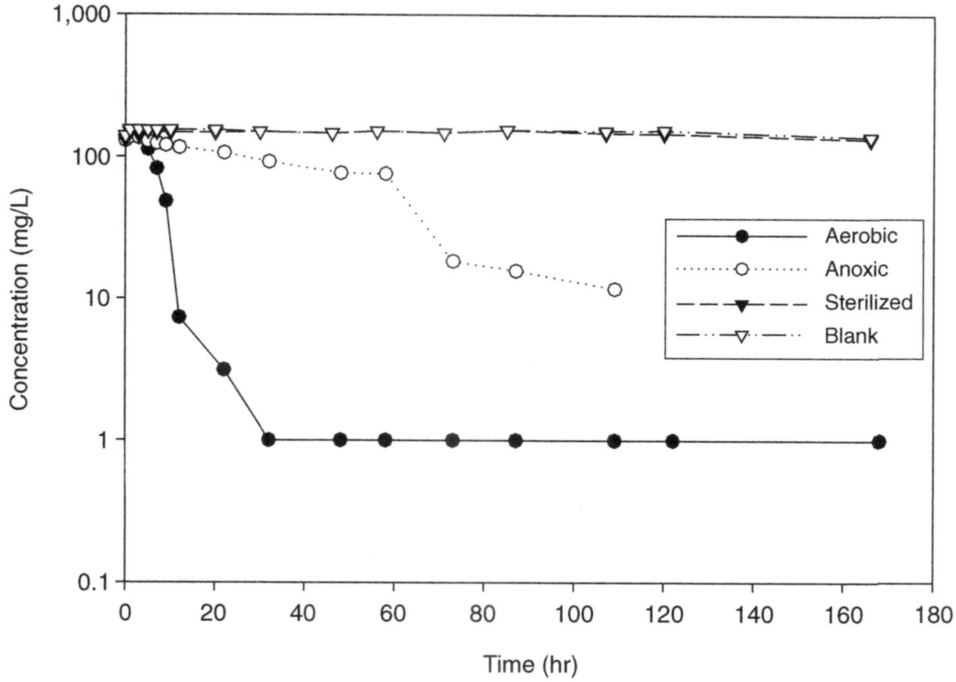

Figure 3.9 BrClAA concentration in batch incubations

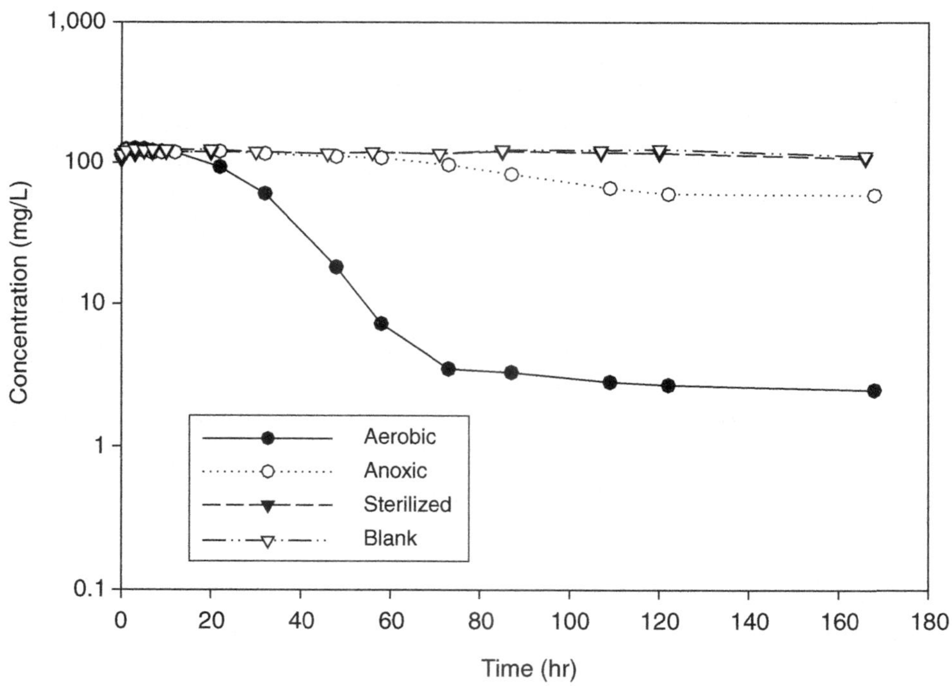

Figure 3.10 Cl_3AA concentration in batch incubations

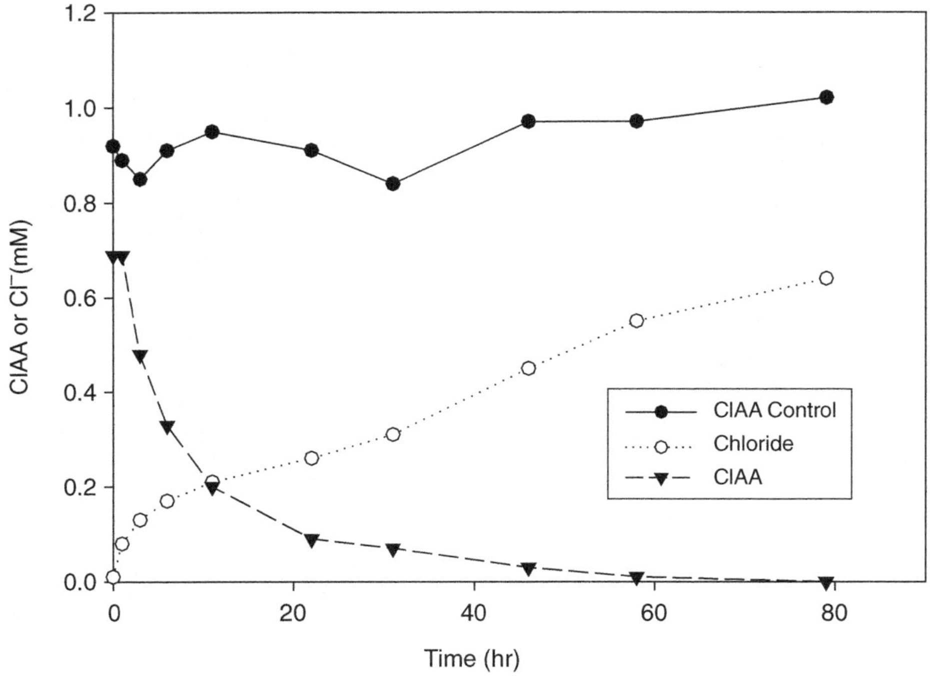

Figure 3.11 ClAA degradation in the presence of BAC

Release of Halide Ions

The release of chloride from BAC incubated with ClAA, Cl_2AA, and Cl_3AA was monitored to confirm the biodegradation (Figures 3.11–3.13). In the first 20 hours, the chloride release rate did not correspond with the rapid HAA decrease for all three HAAs. This may be due to the high HAA initial concentration used (~1 mM). The BAC was collected from a drinking water filter where average Cl_2AA and Cl_3AA levels in the aqueous phase were less than 1 μM. Re-equilibrium of HAA concentrations between GAC and the aqueous phase initiated the HAA adsorption in the beginning of the incubation and reduced the aqueous concentration of HAAs available for biodegradation. By the mass balance, ClAA was close to being completely dechlorinated. If we assume that biodegradation releases all the chloride in Cl_2AA and Cl_3AA molecules, approximately 60% of Cl_2AA and 15% of Cl_3AA were degraded during the incubation period. This indicates that the ease of biodegradation is probably ClAA > Cl_2AA > Cl_3AA. The actual number or series of chloride releases from each HAA molecule is unclear and requires further investigation.

CONCLUSIONS

The biofilm activity inhibition factors, including thermal stress (4°C and 50°C) and chloramine shock, reduced the HAA removal rates in BAC columns significantly. Regarding the respiration characteristics of the HAA-degrading bacteria, an anoxic environment did not affect the HAA removal in column tests. In the batch test, all HAAs except Cl_3AA degraded under

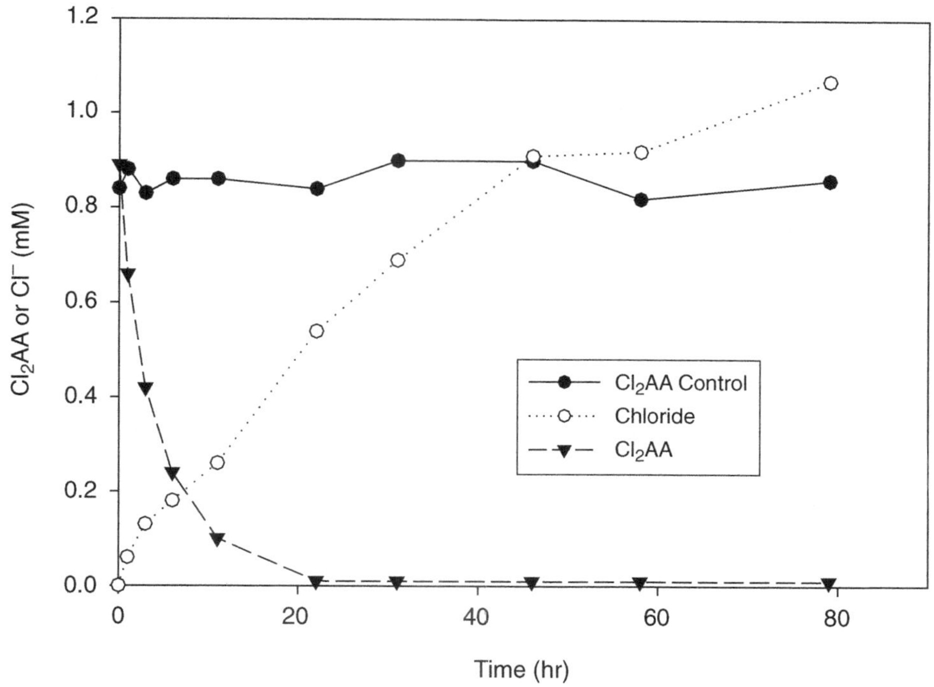

Figure 3.12 Cl_2AA degradation in the presence of BAC

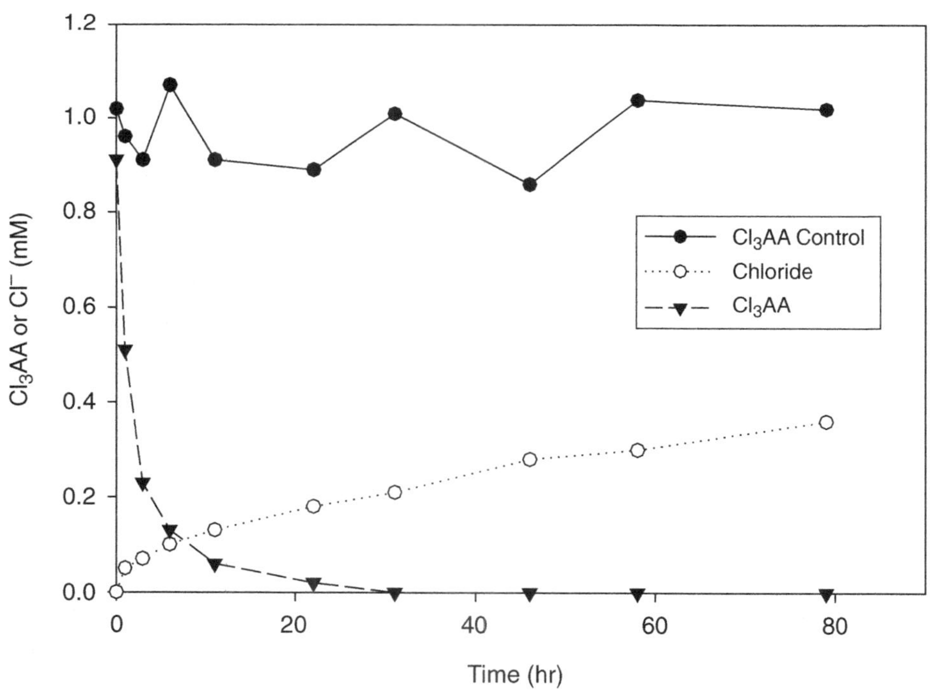

Figure 3.13 Cl_3AA degradation in the presence of BAC

both aerobic and anoxic conditions. For HAA degradation, however, a lag phase was observed in anoxic conditions. Chloride release from HAAs incubated with BAC for 80 hours indicated complete dehalogenation for ClAA, 60% dehalogenation for Cl_2AA, and 15% dehalogenation for Cl_3AA.

These results indicate that biological dehalogenation, under both aerobic and anoxic conditions, plays a major role in HAA removal in BAC filters.

CHAPTER 4
EFFECTS OF EMPTY-BED CONTACT TIME AND TEMPERATURE

INTRODUCTION

In an earlier study, Xie and Zhou (2002) found that HAAs can be effectively removed in a BAC column operated with an EBCT of 20 minutes at room temperature. Their results indicated that biodegradation played a major role in HAA removal. Similar results on HAA biodegradation were also reported by Singer et al. (1999, 2002). Their studies indicated that the removal of HAAs, presumably by biodegradation processes within the GAC filter bed, was more effective when water temperatures were high. In another study, Wobma et al. (2000) reported that HAAs were easily removed in GAC filters through biodegradation. The GAC filters were operated with an EBCT that ranged from 3.6 minutes to 8.4 minutes. A greater HAA removal was observed in June and July versus March, April, and May, pointing to the potential effects of temperature.

Average EBCTs for GAC filters range from 5 to 20 minutes. The question is whether HAAs are effectively removed at these EBCTs. Further, as water temperature changes in various seasons, HAA removal at low water temperatures needs to be investigated.

The objectives of this study were to investigate the effects of EBCT and water temperature on HAA removal in BAC columns and explore kinetic models for HAA removal at various EBCTs and temperatures.

MATERIALS AND METHODS

BAC and water samples were collected from a local surface water treatment plant. The plant employs rapid mixing, flocculation, clarification, GAC–sand filtration, and intermediate and final chlorination. The depth of the GAC medium in the filter is about 106.7 cm (42 in.); the average EBCT for the GAC filter is approximately 16 minutes. BAC samples were obtained from a GAC filter in this plant, which had been online for approximately 2½ years. The BAC sample was collected when the raw water temperature in the plant was near the study temperatures in order to minimize the time required to equilibrate BAC columns.

Four BAC columns were packed with BAC using 50-mL glass burettes and placed in a temperature-controlled incubator. The EBCT of BAC columns was controlled by adjusting the water flow rate. The experiments were conducted at four different temperatures (4°, 10°, 20°, and 30°C) and four different EBCTs (5, 10, 15, and 20 minutes). Finished water from the local water treatment plant was spiked with HAA6 at 50 μg/L each and used as the BAC influent. BAC influent and effluent samples were collected hourly through an 8-hour study period, which started after a 12-hour equilibration. The equilibration of BAC columns was achieved by running the synthetic influent at the study temperature and EBCT for 12 hours.

HAA concentrations in BAC influents and effluents were determined using USEPA Method 552.2 (USEPA 1995a) with a few modifications (Xie, Reckhow, and Springborg 1998). The analytical procedure involves water sample extraction with MTBE, acidic methanol derivatization, and analysis with a GC–ECD. Before sample extraction, HAA samples were preserved with sulfuric acid (pH <2.0) and kept at 4°C.

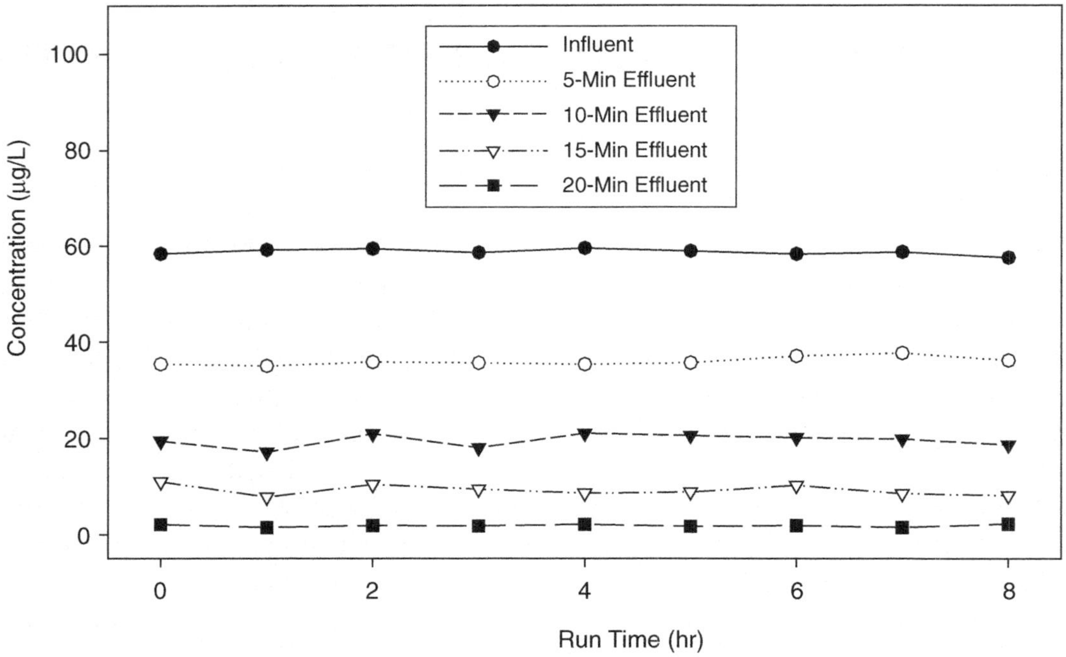

Source: Wu and Xie 2003.

Figure 4.1 Influent and effluent concentrations of ClAA at 4°C

RESULTS AND DISCUSSION

Effects of EBCT on HAA Removal

A typical set of steady-state influent and effluent ClAA concentrations at 4°C and four different EBCTs are shown in Figure 4.1. The influent concentration was around 60 µg/L. Effluent concentrations with EBCTs at 5, 10, 15, and 20 minutes were 35 µg/L, 19 µg/L, 9 µg/L, and 2 µg/L, respectively. The results indicate that increasing the EBCT of the BAC filter increases the removal of ClAA in the BAC filter.

The removal of six HAAs at 4°C is summarized in Figure 4.2. The average ClAA removal efficiency increased from 39% at an EBCT of 5 minutes to 67% at 10 minutes, 84% at 15 minutes, and 97% at 20 minutes. The removal for Cl_2AA increased from 27% at an EBCT of 5 minutes to 55% at 10 minutes, 69% at 15 minutes, and 89% at 20 minutes. The removal efficiency for BrClAA and Br_2AA was comparable with that for Cl_2AA. The removal for Cl_3AA increased from 7% at an EBCT of 5 minutes to 26% at 10 minutes, 35% at 15 minutes, and 65% at 20 minutes. At the same EBCT, the removal of Cl_3AA was the lowest while those of ClAA and BrAA were the highest. This indicates that HAA removal decreases as its halogen number increases, corroborating the results obtained from Zhou and Xie's (2002) study.

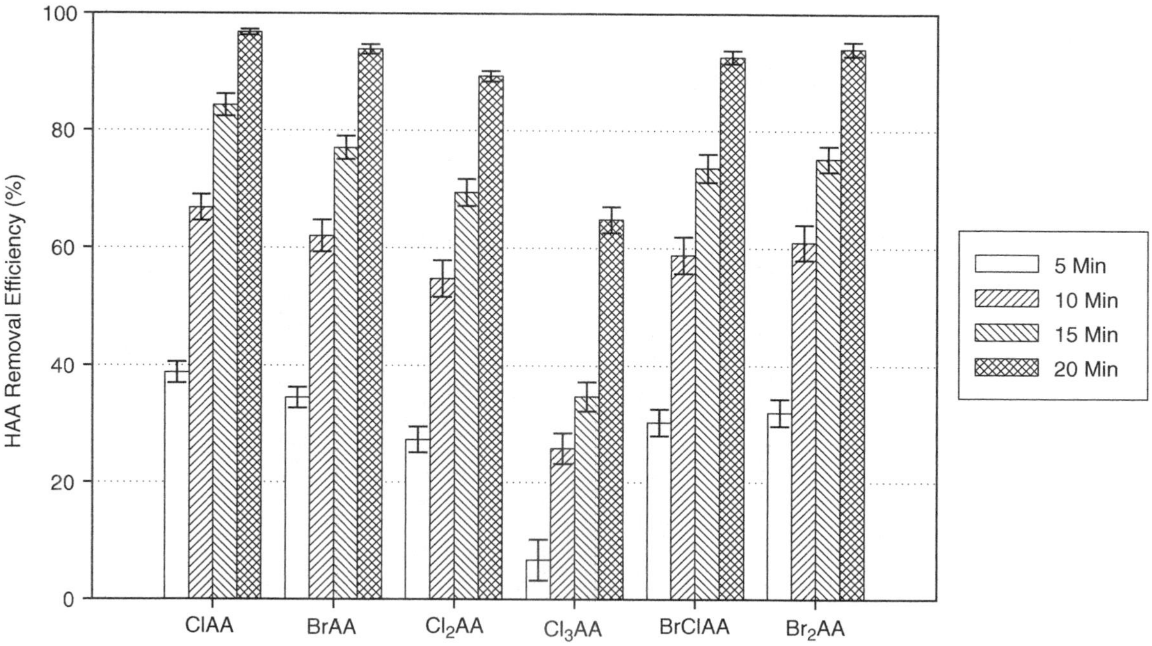

Source: Wu and Xie 2003.

Figure 4.2 HAA removals at 4°C

The average removal efficiency for the six HAAs was 28% at an EBCT of 5 minutes. By increasing the EBCT to 10, 15, and 20 minutes, the average removal efficiencies for the six HAAs increased to 55%, 69%, and 89%, respectively. The results indicate that EBCT significantly affected the HAA removal in BAC columns, and raising the EBCT increased the HAA removal. Therefore, an EBCT of 10 minutes may be needed to obtain a 50% removal of HAAs in low-temperature (4°C) waters.

The removal of six HAAs at 10°C is summarized in Figure 4.3. ClAA removal efficiency increased from 80% at an EBCT of 5 minutes to 98% at 10 minutes, 99% at 15 minutes, and almost 100% at 20 minutes. The removals of BrAA, Cl_2AA, BrClAA, and Br_2AA were slightly lower than that of ClAA. The average removal efficiency of all monohalo- and dihaloacetic acids was 65%, 94%, 98%, and 99% at EBCTs of 5, 10, 15, and 20 minutes, respectively. Therefore, a near complete removal was achieved for these five HAA species in the BAC column at 10°C and an EBCT of 10 minutes. Further increasing EBCT did not significantly benefit the removal. The removal efficiency for Cl_3AA increased from 23% at an EBCT of 5 minutes to 50% at 10 minutes, 67% at 15 minutes, and 79% at 20 minutes. This removal is less than that for mono- and dihalo-acetic acids. The average removal efficiencies for HAA6 were 58%, 87%, 93%, and 95% at EBCTs of 5, 10, 15, and 20 minutes, respectively. Fifty-eight percent of HAA removal was achieved even at the lowest EBCT, 5 minutes. Increasing the EBCT increased HAA removal.

At 20°C and 30°C, monohalo- and dihaloacetic acids were almost completely removed for all four EBCTs, as shown in Figures 4.4 and 4.5. The removal efficiency for Cl_3AA at 20°C increased from 56% at an EBCT of 5 minutes to 80% at 10 minutes, 92% at 15 minutes, and 96% at 20 minutes. At 30°C, the lowest removal efficiency of Cl_3AA was 88% at an EBCT of 5 minutes. At an EBCT of 10 minutes or higher, a nearly complete removal of Cl_3AA was

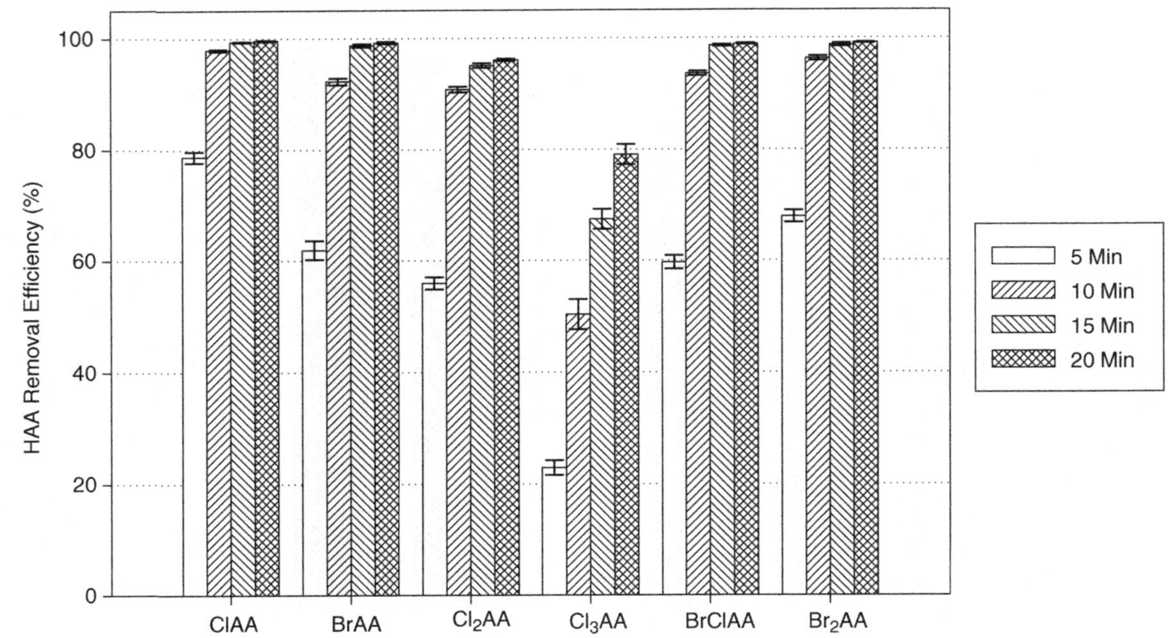

Source: Wu and Xie 2003.

Figure 4.3 HAA removals at 10°C

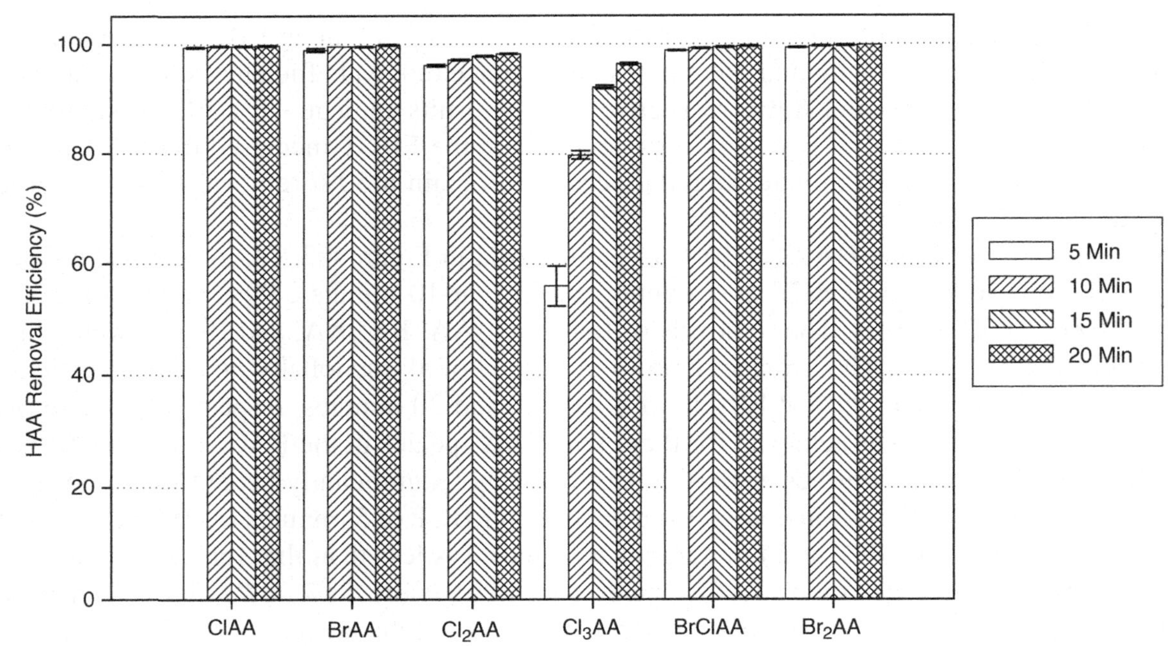

Source: Wu and Xie 2003.

Figure 4.4 HAA removal at 20°C

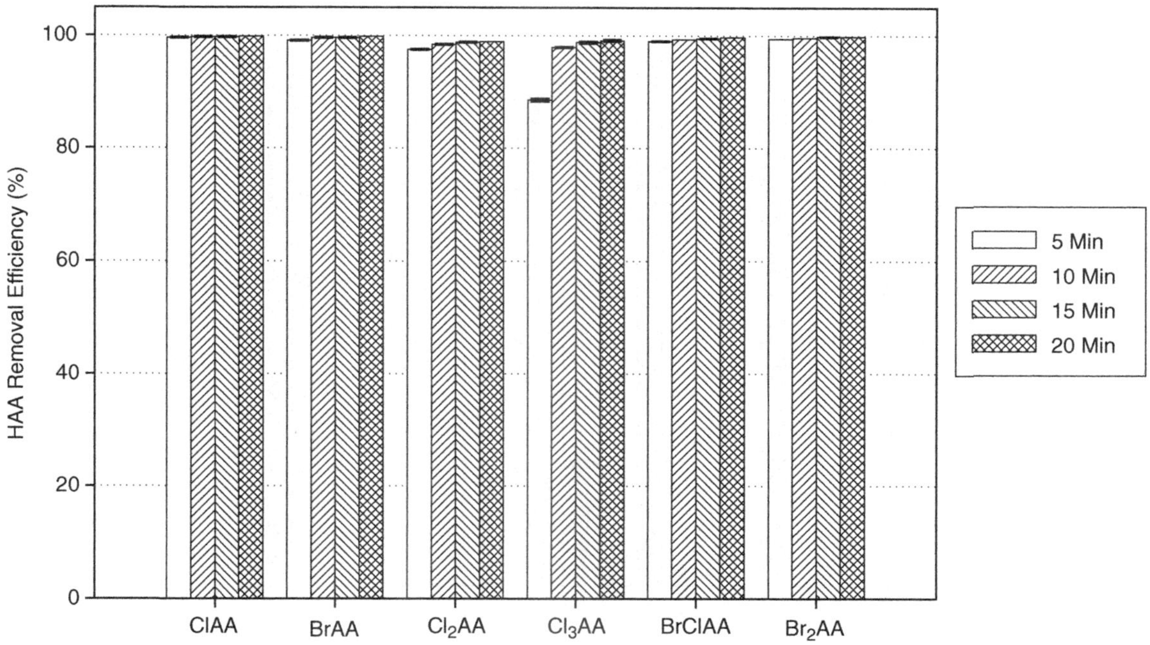

Source: Wu and Xie 2003.

Figure 4.5 HAA removal at 30°C

achieved. The average removal efficiency of six HAAs at 20°C was 91%, 96%, 98%, and 99% at EBCTs of 5, 10, 15, and 20 minutes, respectively. At 30°C, the removal efficiencies were 97%, 99.1%, 99.4%, and 99.5% at EBCTs of 5, 10, 15, and 20 minutes, respectively. At 20°C and 30°C, a nearly complete removal of HAAs (except Cl_3AA) in BAC columns was obtained even with the lowest EBCT: 5 minutes. Further increasing EBCT had little benefit on HAA removal.

Effects of Water Temperature on HAA Removal

HAA removal efficiencies at four water temperatures are shown in Figures 4.6–4.9. Figure 4.6 illustrates HAA removal at an EBCT of 5 minutes. The removal for Cl_2AA increased from 27% at 4°C to 56% at 10°C, 96% at 20°C, and 98% at 30°C. The removals for ClAA, BrAA, BrClAA, and Br_2AA were comparable with that of Cl_2AA at the four temperatures.

Cl_3AA removal efficiency was 7% at 4°C, 23% at 10°C, 56% at 20°C, and 88% at 30°C. The average removal efficiency of HAA6 was 28%, 58%, 91%, and 97% at 4°, 10°, 20°, and 30°C, respectively. The results indicate that water temperature significantly affected HAA removal in BAC columns. Increasing the temperature increased the removal efficiency of HAAs. BAC filtration with an EBCT of 5 minutes can remove 58% of HAAs at a water temperature of 10°C or higher.

At an EBCT of 10 minutes, increasing the temperature also increased the removal efficiency of HAAs, as shown in Figure 4.7. The results indicate that BAC filtration with an EBCT of 10 minutes can remove 55% of HAAs, even at 4°C water. When the EBCT increased to 15 and 20 minutes, HAAs (except Cl_3AA) were almost completely removed at 10°, 20°, and 30°C. At an EBCT of 15 minutes, Cl_3AA removal efficiency increased from 35% at 4°C to 67% at 10°C, 92%

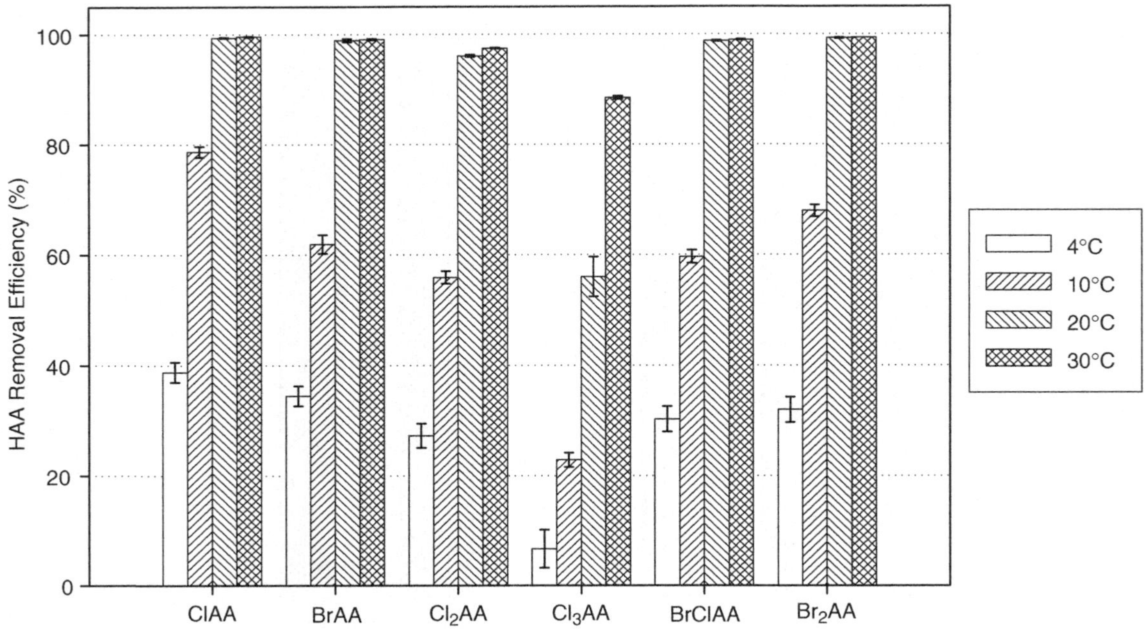

Source: Wu and Xie 2003.

Figure 4.6 HAA removal at an EBCT of 5 minutes

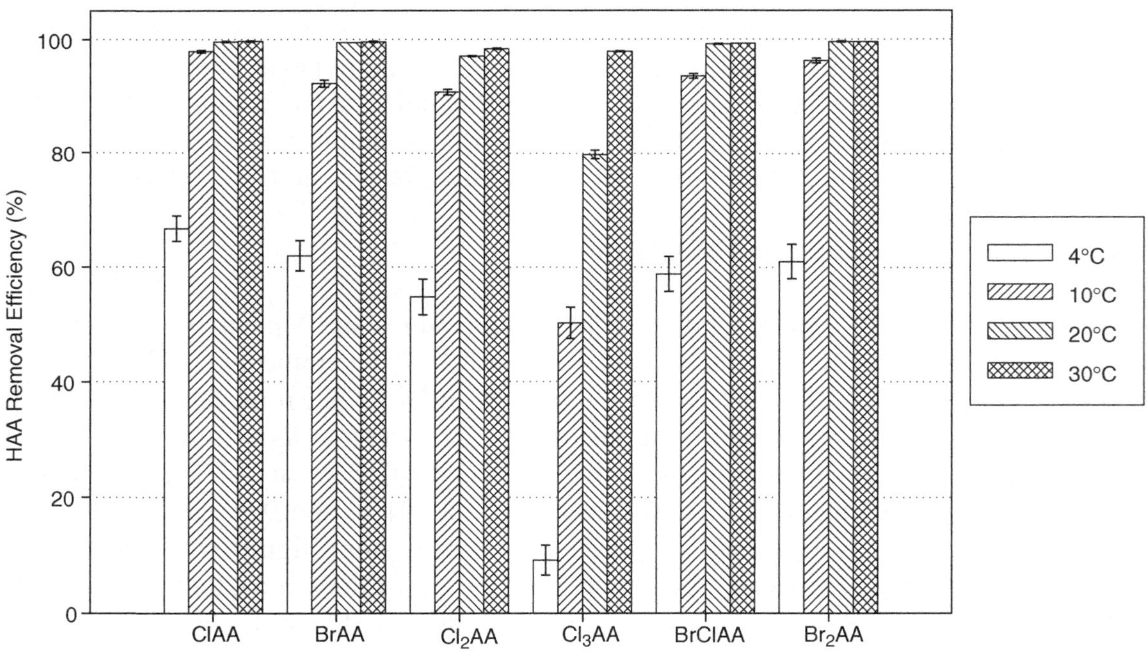

Source: Wu and Xie 2003.

Figure 4.7 HAA removal at an EBCT of 10 minutes

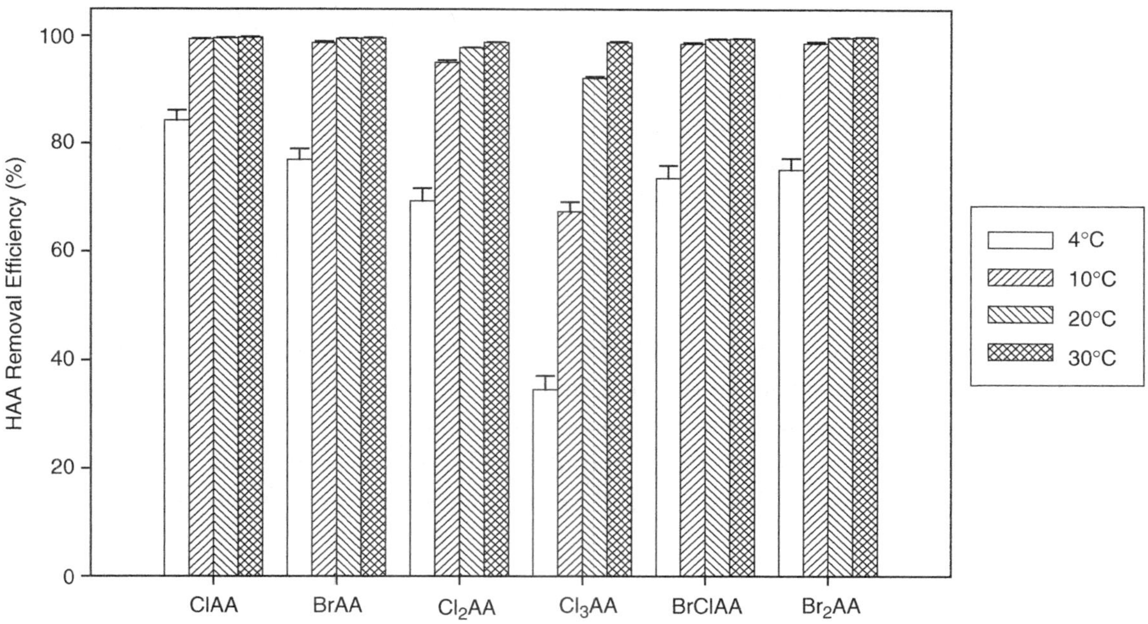

Source: Wu and Xie 2003.

Figure 4.8 HAA removal at an EBCT of 15 minutes

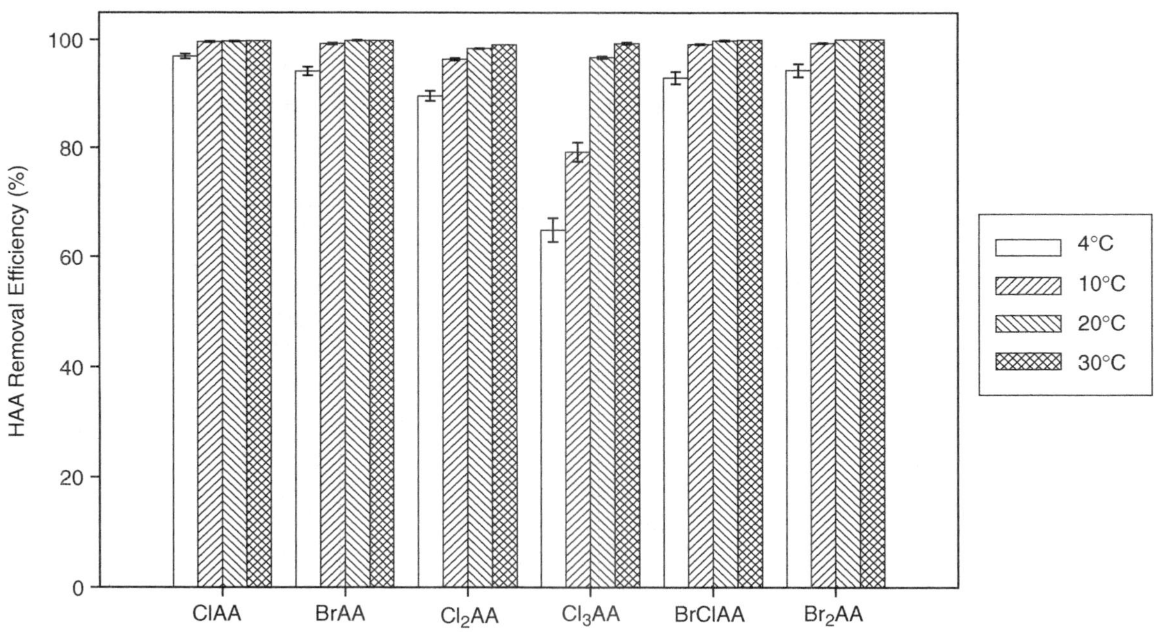

Source: Wu and Xie 2003.

Figure 4.9 HAA removal at an EBCT of 20 minutes

at 20°C, and 99% at 30°C. At an EBCT of 20 minutes, Cl_3AA removal efficiency increased from 65% at 4°C to 79% at 10°C, and 96% at 20°C and 30°C.

In conclusion, both EBCT and water temperature affected the removal of HAAs using BAC filtration. A longer EBCT could be used to compensate for the effect of a low water temperature. For higher than 50% removal of HAA6, the authors suggest a 10-minute EBCT for 4°C waters and a 5-minute EBCT for waters 10°C or higher.

Kinetic Models of HAA Removal

A variable-order kinetic model, $R = kC^n$, is used to describe the HAA removal in BAC filters. In this model, R is the removal rate, k is the reaction rate constant, C is the concentration of a specific reactant species, and n is the reaction order (Logan 2001).

The influent concentrations (C_{in}) and the effluent concentrations (C_{eff}) of the BAC columns obtained at various EBCTs were used to calculate the HAA removal rates, R, as shown in the following equation:

$$R = (C_{\text{in}} - C_{\text{eff}})/\text{EBCT}$$

Because of concentration changes in the BAC columns, log-mean concentration in each BAC column, C_{lm}, was calculated using the following equation:

$$C_{\text{lm}} = (C_{\text{in}} - C_{\text{eff}})/ln(C_{\text{in}}/C_{\text{eff}})$$

By plotting R versus C_{lm} in logarithm scale, a regression line was obtained, as shown in Figure 4.10. The reaction order can be determined by the slope of the regression line.

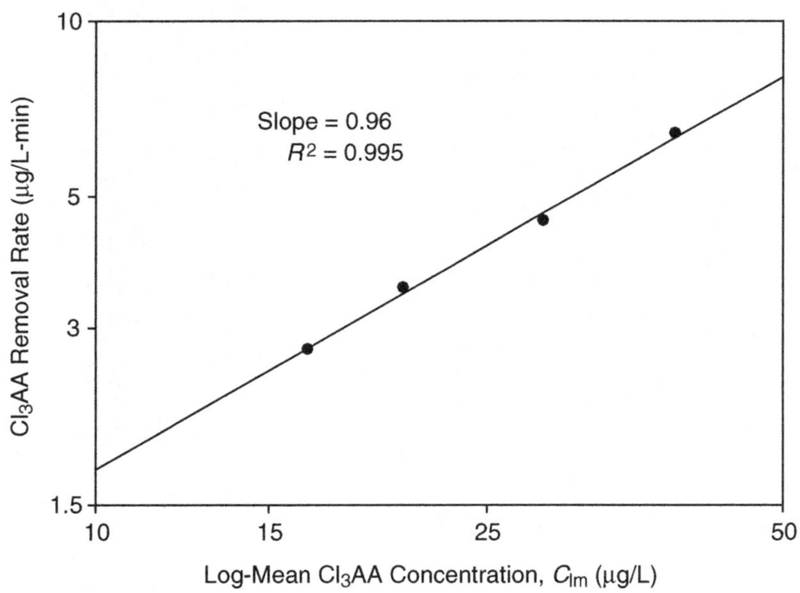

Source: Wu and Xie 2003.

Figure 4.10 Cl_3AA removal rates versus log-mean concentrations at 20°C

Table 4.1
HAA removal reaction orders and correlation coefficients at 4°C

HAA	Slope (n)	R^2
ClAA	0.46	0.97
BrAA	0.45	0.91
Cl_2AA	0.30	0.78
BrClAA	0.34	0.86
Br_2AA	0.36	0.86
Cl_3AA	–1.68	0.72

Source: Wu and Xie 2003.

Table 4.2
HAA removal reaction orders and correlation coefficients at 10°C

HAA	Slope (n)	R^2
ClAA	1.02	0.92
BrAA	0.69	0.97
Cl_2AA	0.86	0.79
BrClAA	0.70	0.86
Br_2AA	0.86	0.86
Cl_3AA	0.29	0.55

Source: Wu and Xie 2003.

Most HAA effluent concentrations at 20°C and 30°C were near their method detection limits. Because those data could not accurately describe HAA removal rates, they were not used for modeling. HAA concentrations obtained at two temperatures (4°C and 10°C) and Cl_3AA concentrations at 20°C were calculated for the removal rate R and C_{lm}. A representative plot on Cl_3AA removal rates versus Cl_3AA concentrations at 20°C is shown in Figure 4.10.

As shown in Table 4.1, the reaction orders of HAA removal at 4°C were below 0.5. This suggests that the HAA removals in the BAC columns at 4°C were likely independent of HAA concentrations and tended to follow the zero-order model. These findings indicate that HAA removal was limited by the chemical or biological properties of BAC columns, such as adsorption capacity and biological activities in BAC columns.

At 10°C, the reaction orders of monohaloacetic acid and dihaloacetic acid removal were close to 1, as shown in Table 4.2. This suggests that their removals at 10°C in the BAC columns were dependent on HAA concentrations and tended to follow the first-order reaction model. This indicates that biological activities in BAC columns were sufficient, and HAA removals were limited by their concentrations. However, the reaction order for Cl_3AA removal at 10°C was 0.29, suggesting that Cl_3AA removal at 10°C was likely independent of its concentration and tended to follow the zero-order model. This indicates that there were limited biological activities for Cl_3AA degradation at 10°C, supporting the finding that Cl_3AA was poorly removed in BAC columns, and a different bacteria species may be involved for Cl_3AA degradation, as discussed in chapter 3. At 20°C, the Cl_3AA removal model followed the first-order model (Figure 4.10). This indicates that sufficient biological activities had been developed and Cl_3AA removal was dependent only on its concentration. The model for Cl_3AA removal at 30°C was not calculated because its concentrations were near the detection limits and therefore not reliable.

CONCLUSIONS

Results of this study indicate that both EBCT and water temperature significantly affected HAA removal in BAC columns. Raising the water temperature and EBCT increased the removal of HAAs. At a temperature at 20°C or higher, increasing EBCT had little effect on HAA removal. A longer EBCT could be used to compensate for the effect of a low water temperature. For an HAA6 removal efficiency at 50% or higher, the authors suggest a 10-minute EBCT for 4°C waters and a 5-minute EBCT for waters at 10°C or higher.

HAA removal in BAC columns had different kinetic models at different temperatures. At 4°C, HAA removal tended to follow a zero-order reaction model. This indicates that HAA removal may be limited by carbon adsorption capacity or biological activities on its surface. At 10°C, the monohaloacetic acid and dihaloacetic acid removal tended to follow a first-order reaction model. This indicates that biological activities were fully developed at 10°C and HAAs removal was limited by their concentrations. For Cl_3AA removal, a first-order reaction model was obtained at 20°C but not at 10°C.

CHAPTER 5
HAA REMOVAL IN FULL-SCALE GAC FILTERS

INTRODUCTION

Recent studies indicate that HAAs undergo biological degradation in drinking water (Singer et al. 1993; Williams, Rindfleisch, and Williams 1995; Williams, Williams, and Gordon 1996; Thomas et al. 2000; Zhou and Xie 2002). Zhou and Xie (2002) suggested that biodegradation could be used as a treatment technology to lower HAA levels in finished water. In another study, Xie and Zhou (2002) found that HAAs were effectively removed in a BAC column with an EBCT of 20 minutes. Similar results were also reported by Singer et al. (1999) and Singer et al. (2002). Their studies indicated that the removal of HAAs, presumably by biodegradation processes within the GAC filter, was more effective when water temperatures were high. Wobma et al. (2000) also reported that HAAs were easily removed in GAC filters through biodegradation. The GAC filters were operated with EBCTs ranging from 3.6 minutes to 8.4 minutes.

Because of biological activities on the surface of GAC, GAC filtration could be an effective treatment process for HAA removal. In this task, HAA removal using GAC filters was investigated in four full-scale water treatment plants. The effect of GAC filtration on the HAA levels in six distribution systems was evaluated as well.

MATERIALS AND METHODS

This full-scale GAC filter study was conducted at three water treatment plants in Pennsylvania and one water treatment plant in California. The design and operation parameters of GAC filters in all four water treatment plants are listed in Table 5.1. During the study, each plant had parallel GAC filters with different online ages. Both influent and effluent samples for the selected filters were periodically collected at the plants.

Table 5.1
Design and operation parameters of GAC filters in four water treatment plants

	Plant A		Plant B		Plant C		Plant D	
GAC	Filter 1	Filter 5	Filter 1	Filter 2	Filter 4	Filter 5	Filter 1	Filter 2
Depth, cm (in.)	106.7 (42)	106.7 (42)	106.7 (42)	106.7 (42)	71.1 (28)	50.8 (20)	111.8 (44)	111.8 (44)
EBCT* (min)	16.3	16.3	8.8	8.8	13.6	5.1	9.2	9.2
Replacement date	9/16/99 and 10/18/02	2/22/01	3/3/02	9/11/01	5/30/02	11/13/02	3/22/02	Late 1995

*Based on the average flow in 2002 and the number of GAC filters.

Six distribution systems in Pennsylvania were chosen for the study. Three of the systems received water treated with GAC filtration and three without GAC filtration. Prechlorination or intermediate chlorination was practiced in these systems. Samples were collected before and after GAC filtration; at the entry to distribution systems; and at 25%, 50%, and maximum residence time locations in a cold month (April) and a warm month (August).

HAA concentrations were determined using USEPA Method 552.2 (USEPA 1995a) with a few modifications (Xie, Reckhow, and Springborg 1998). The procedure involves water sample extraction with MTBE, acidic methanol derivatization, and analysis with a GC–ECD. Before sample extraction, HAA samples were preserved with ammonium chloride and sulfuric acid (pH <2.0) and kept at 4°C. THM concentrations were analyzed using a modified USEPA Method 551.1 (USEPA 1995b), which involves water sample extraction with MTBE and analysis with a GC–ECD. Before sample extraction, THM samples were preserved with ammonium chloride and phosphate buffer and kept at 4°C.

RESULTS AND DISCUSSION

HAA Removal in Field GAC Filters

Plant A

Plant A is a conventional treatment plant with coagulation, floc-blanket sedimentation, and GAC–sand filtration. Influent and effluent samples were collected monthly from two parallel GAC filters—1 and 5—from December 2001 to February 2003, for a total of 15 months.

A significant reduction in HAA concentrations was observed in both filters during the study. HAA6 removal efficiency ranged from 37% to 97%, with an average of 76% in filter 1 and 79% in filter 5. The influent and effluent concentrations of two predominant HAAs—Cl_2AA and Cl_3AA—for the two GAC filters in plant A are shown in Figures 5.1 and 5.2.

The Cl_2AA removal efficiency in the two filters ranged from 52% to 98%. For filter 1, a 52% reduction occurred in January 2002, which was the coldest month. The average removal for filter 1 in winter (from December to February) was 71%. This indicates that a major reduction of Cl_2AA was achieved inside the GAC filter, even at low water temperatures. Although the GAC in filter 1 and filter 5 had different online ages (approximately 1½ years' difference), the filters' average removal efficiencies for Cl_2AA, 80% and 84%, were comparable.

The Cl_3AA removal efficiency ranged from 0% to 98% with an average of 64% for filter 1 and 68% for filter 5. The lowest Cl_3AA removal occurred during the coldest months (January and February 2003), and the highest removal occurred right after the GAC was replaced. The average removal for filter 1 was only 20% in winter. When the GAC was newly replaced or at high water temperatures, the Cl_3AA removal for filter 1 and filter 5 was 86% and 83%, respectively.

Limited results on THM reduction in the two GAC filters are shown in Figure 5.3. Before the GAC change, there were no reductions in either filter 1 or filter 5. Right after the GAC change in filter 1, a nearly complete removal of $CHCl_3$, the main THM species, was observed. However, a higher concentration of $CHCl_3$ was observed in the effluent than in the influent after 3 months' operation. This may be due to $CHCl_3$ desorption from the GAC surface.

The results indicate that both new and old GAC filters can remove HAAs. The newly replaced GAC filter had a slightly better removal rate of HAAs for 1 or 2 months. After 2 or 3 months, the carbon online age had little effect on the removal of HAAs. Water temperature had

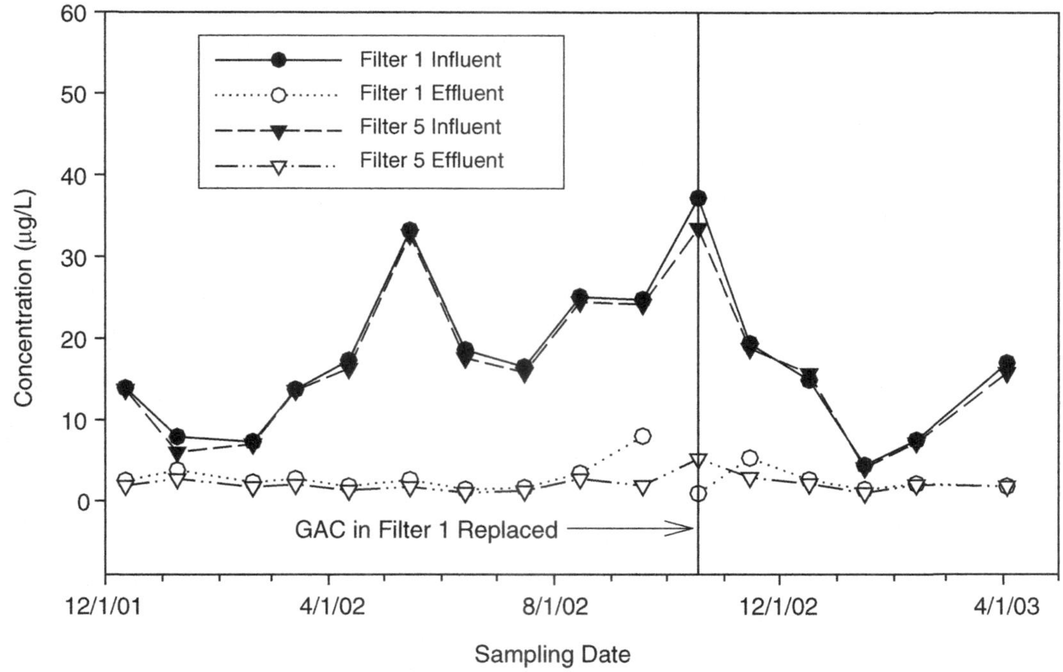

Figure 5.1 Cl_2AA influent and effluent concentrations for GAC filters in plant A

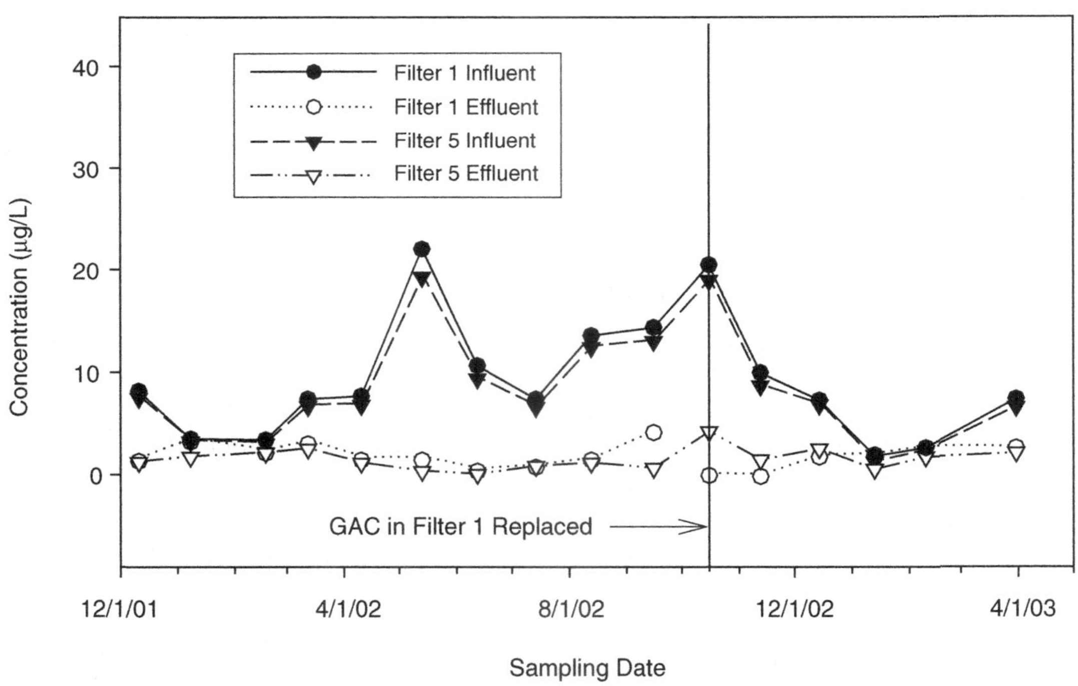

Figure 5.2 Cl_3AA influent and effluent concentrations for GAC filters in plant A

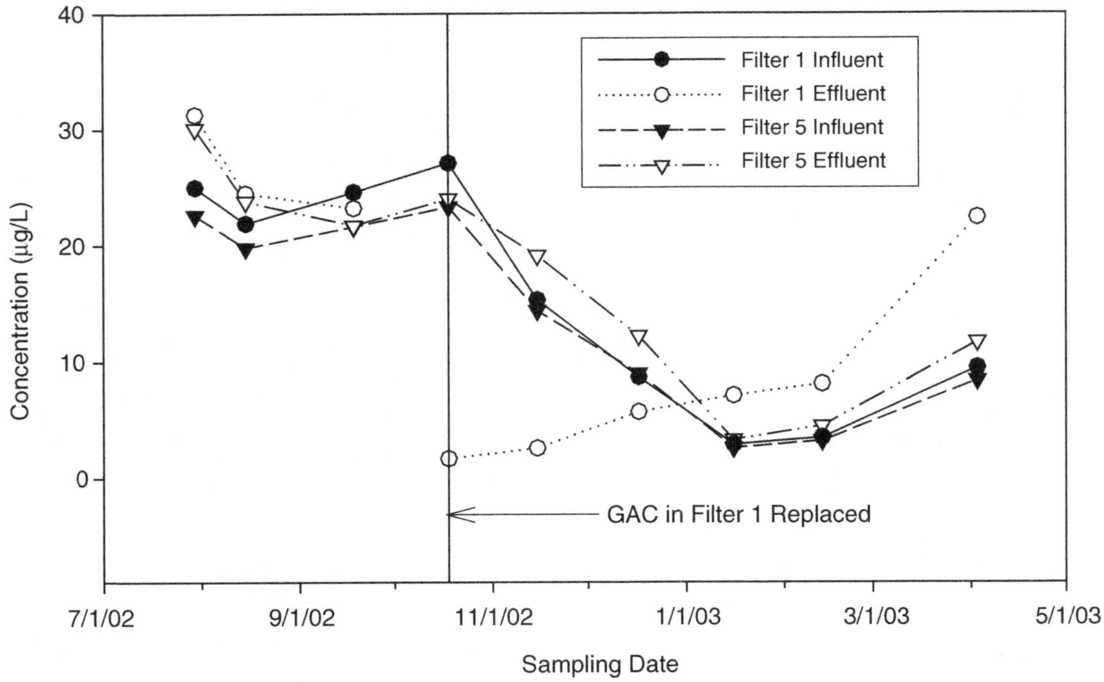

Figure 5.3 $CHCl_3$ influent and effluent concentrations for GAC filters in plant A

a greater impact on the removal of Cl_3AA than on Cl_2AA. The new GAC filter significantly reduced the level of THMs and Cl_3AA for 1 or 2 months. No removal of THMs was observed in aged GAC filters.

Plant B

Plant B has two identical parallel treatment trains after coagulation. Each includes floc-blanket sedimentation and GAC–sand filtration. Influent and effluent water samples were collected monthly for two GAC filters—1 and 2—from December 2001 to February 2003.

An effective reduction of HAAs was found in both GAC filters. The HAA6 removal efficiency ranged from 18% to 91%, with an average removal efficiency of 60% for filter 1 and 63% for filter 2 during the 15-month study. The influent and effluent concentrations for Cl_2AA and Cl_3AA for filter 1 and filter 2 are shown in Figures 5.4 and 5.5.

Cl_2AA removal efficiency ranged from 27% to 94%, with an average value of 70% for filter 1 and 76% for filter 2. The average Cl_2AA removal efficiency in filter 1 in winter (January–March) was 46%. The biggest difference between the two filters in removing Cl_2AA occurred in March 2002, right after the GAC in filter 1 was replaced. The lower removal for Cl_2AA for filter 1 may be due to saturation of Cl_2AA on the GAC and limited bioactivity in the GAC filter 1 after the carbon change. In other months, Cl_2AA removal efficiencies for filter 1 and filter 2 were comparable.

Cl_3AA removal efficiency ranged from 0% to 81% with an average value of 40% for filter 1 and 41% for filter 2. In the coldest months, no Cl_3AA removal was observed in either filter. The

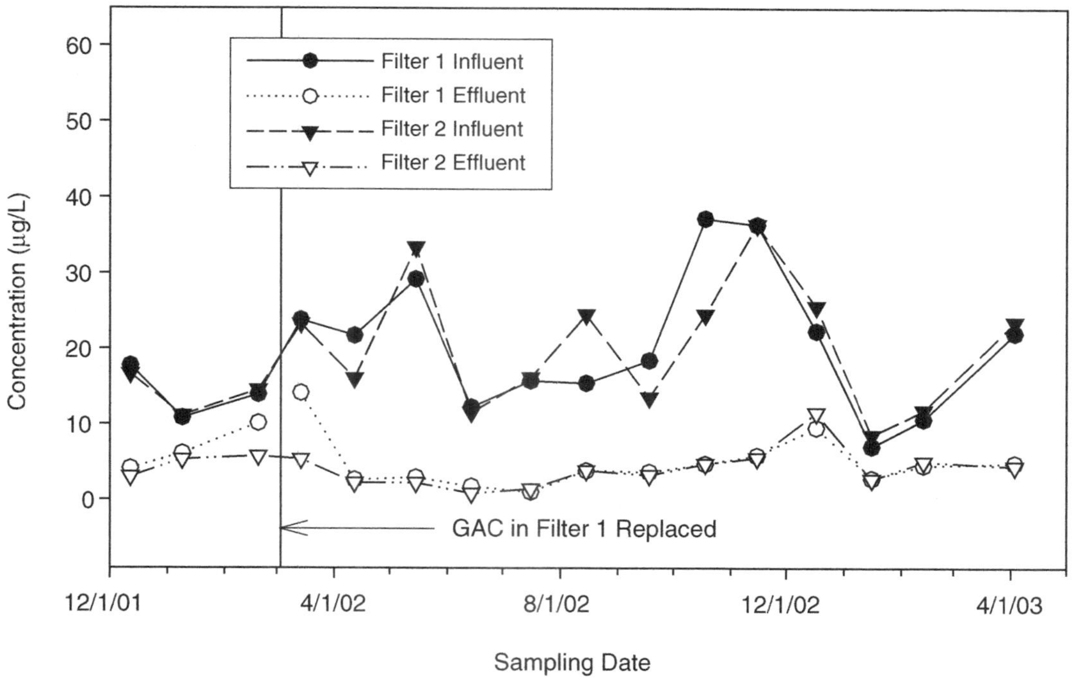

Figure 5.4 Cl_2AA influent and effluent concentrations for GAC filters in plant B

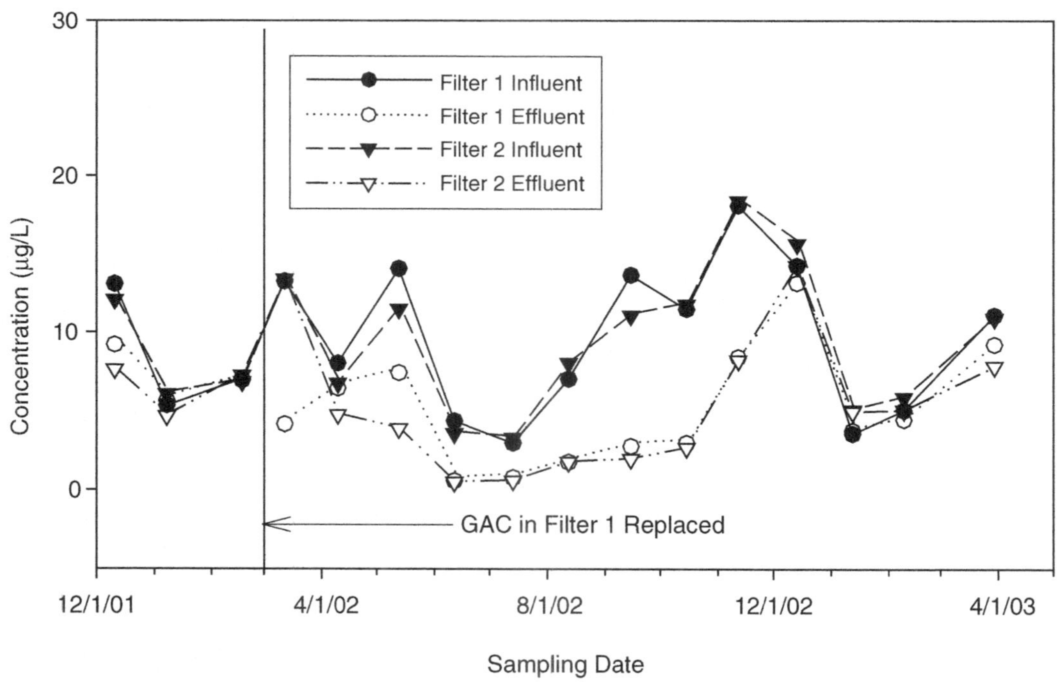

Figure 5.5 Cl_3AA influent and effluent concentrations for GAC filters in plant B

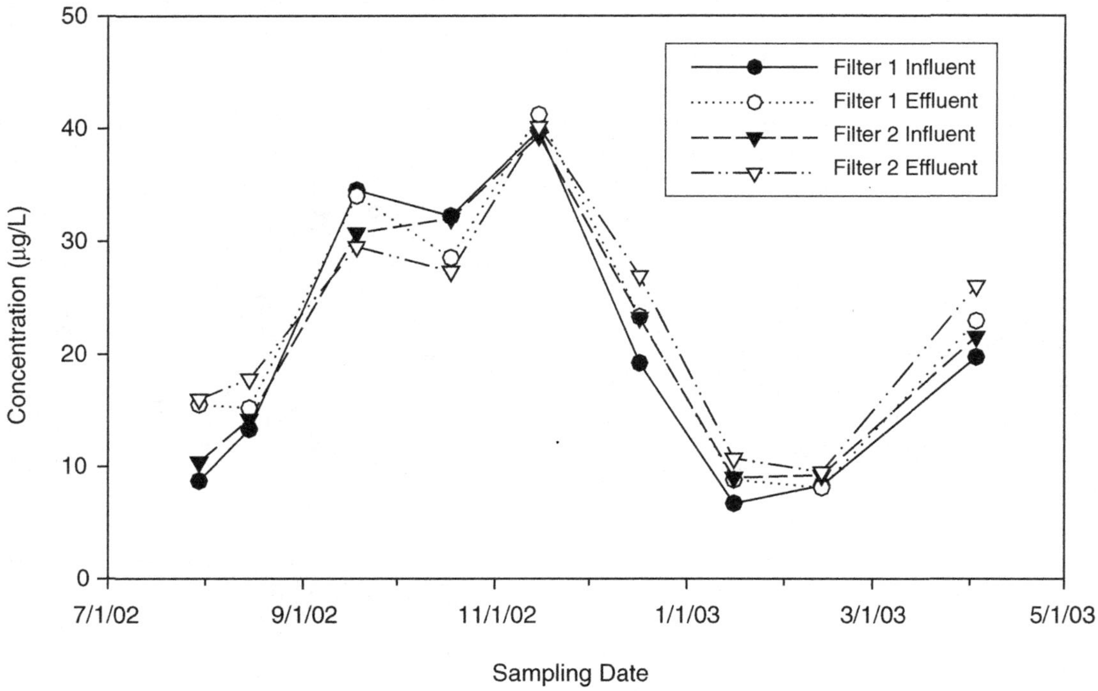

Figure 5.6 $CHCl_3$ influent and effluent concentrations for GAC filters in plant B

highest removal, 81%, occurred during the warm months. This indicates that water temperature had a significant impact on Cl_3AA removal.

A limited study on THM reduction in GAC filters was conducted in plant B. No GAC change occurred during the THM study period. For $CHCl_3$, the predominant THM, there were no reductions in either filter 1 or filter 2, as shown in Figure 5.6.

HAA removal was lower in plant B than in plant A. The better HAA removal efficiency in plant A may be due to its longer EBCT (16.3 minutes) for GAC filters. In addition, the fluctuation of HAA removal efficiency caused by water temperature change in plant B was greater than that in plant A. This indicates that water temperature had a more significant impact on GAC filters with a shorter EBCT (8.8 minutes).

Plant C

Plant C has two different treatment process trains. Train 1 includes coagulation, sedimentation, and GAC–sand filtration. The treatment process in train 2 includes coagulation, floc-blanket sedimentation, and GAC–sand filtration. Influent and effluent water samples were collected monthly from filter 4 in train 1 and filter 5 in train 2 from December 2001 to April 2003.

The removal efficiency of Cl_2AA for filter 4 ranged from 47% in January to 90% in July, with an average value of 75%, as shown in Figure 5.7. Removal of Cl_2AA for filter 5 fluctuated from 9% in January to 99% in November with an average value of 66%. The average removal efficiency of Cl_3AA was 35% for filter 4 and 41% for filter 5, as shown in Figure 5.8. In January, no Cl_3AA was removed in either filter. For both filters, the highest removals occurred right after the new GAC was put into service. A nearly complete removal of $CHCl_3$ was observed in filter 5 right

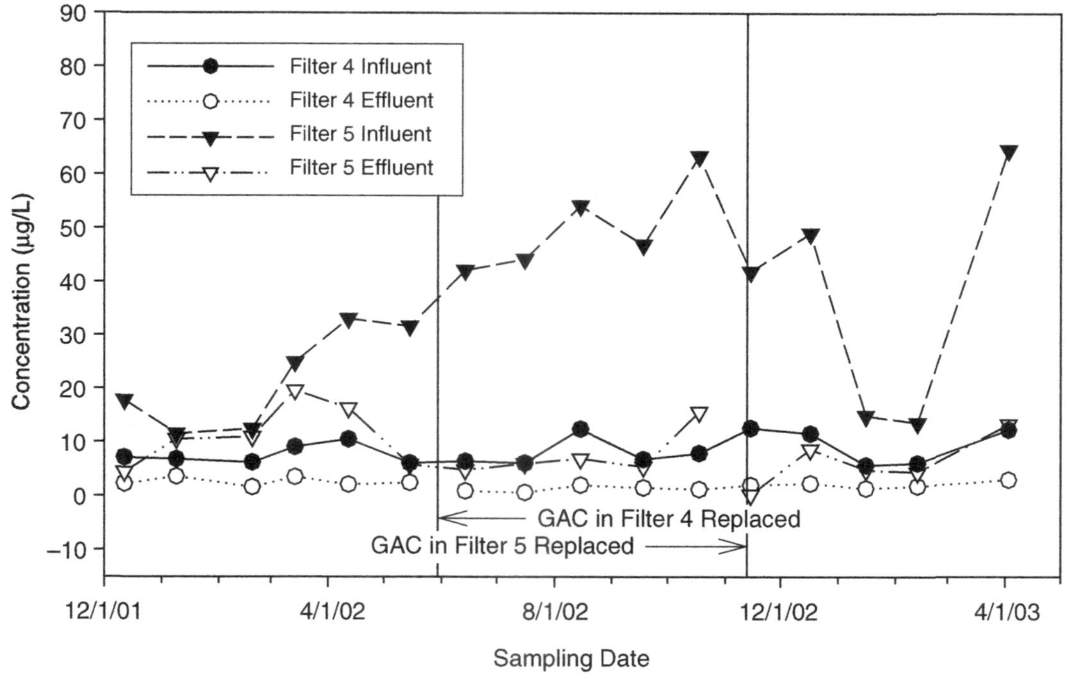

Figure 5.7 Cl_2AA influent and effluent concentrations for GAC filters in plant C

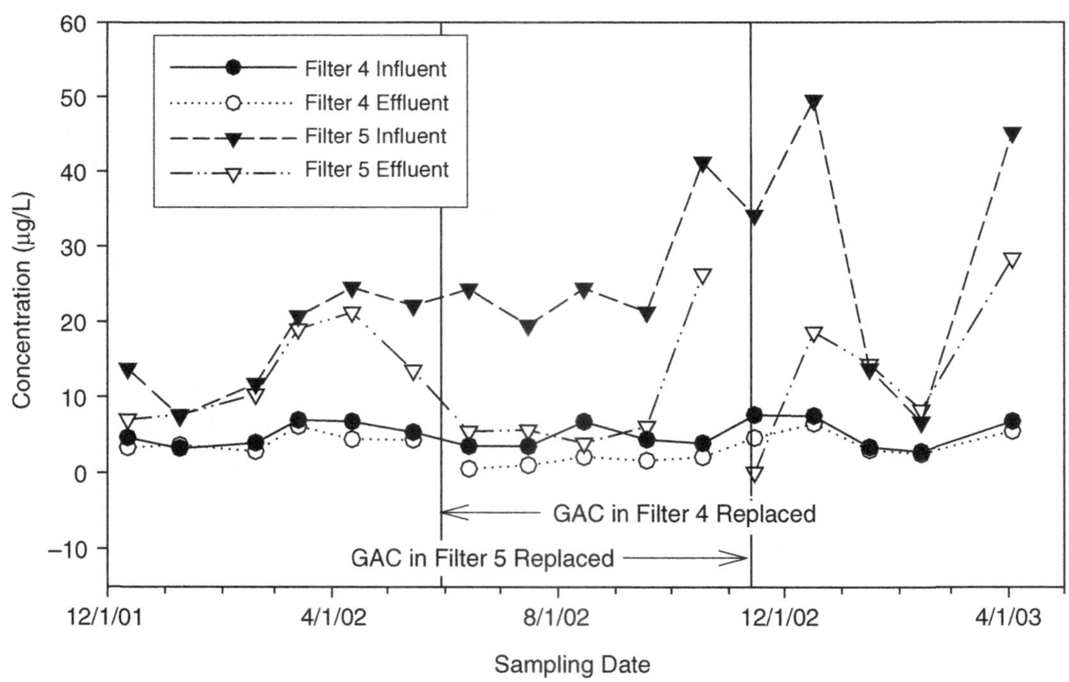

Figure 5.8 Cl_3AA influent and effluent concentrations for GAC filters in plant C

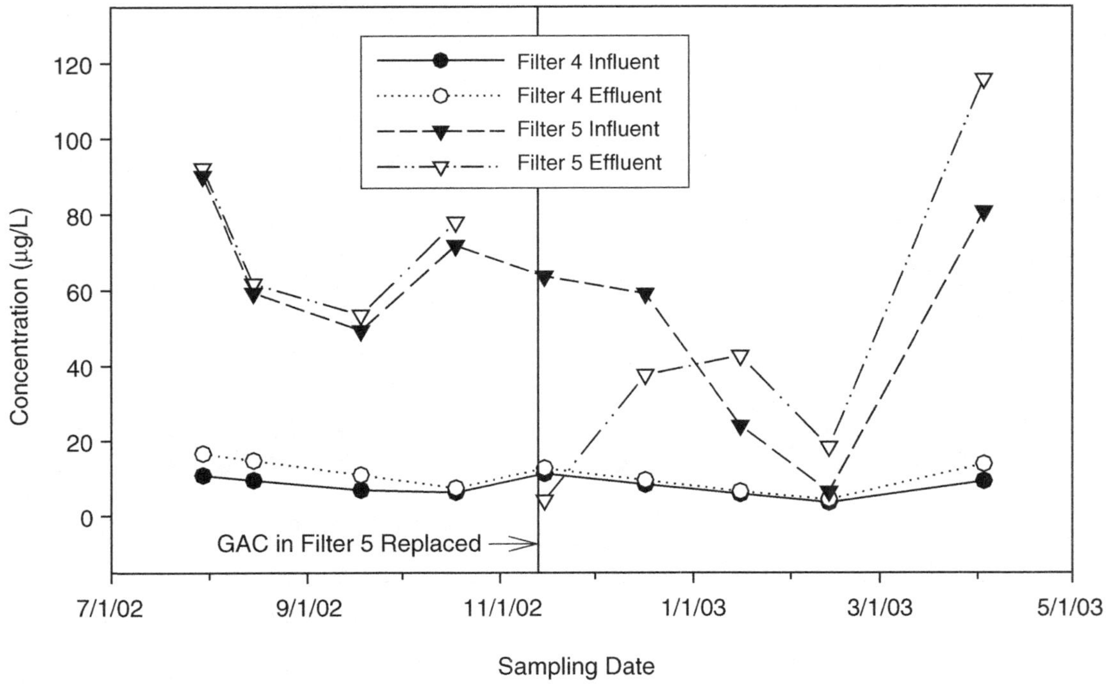

Figure 5.9 $CHCl_3$ influent and effluent concentrations for GAC filters in plant C

after the GAC was replaced, as shown in Figure 5.9. After 2 months' operation, an increase of $CHCl_3$ in the effluent was observed in filter 5 due to the desorption of $CHCl_3$ from the GAC. During other periods, no $CHCl_3$ removal was observed in either filter.

These results were consistent with those obtained in plant A and plant B. The poor HAA removal during the winter season indicates that filter 5 was more susceptible to the cold water temperature effect because of its shorter EBCT (5.1 minutes). A lower average removal for Cl_3AA in filter 4 with a longer EBCT (13.6 minutes) may be due to a much lower Cl_3AA concentration in the influent.

Plant D

Plant D is a conventional water treatment plant with an intermediate ozonation process. Treatment processes include coagulation, sedimentation, intermediate ozonation, and GAC–sand filtration. Influent and effluent water samples from filters 1 and 2 were collected weekly from April 9, 2002, when the carbon was replaced in filter 1, to October 15, 2002, and then monthly until February 2003.

A significant HAA reduction was achieved in both GAC filters, as shown in Figures 5.10, 5.11, and 5.12. HAA6 removal efficiency ranged from 56% to 93%, with an average value of 85% in both filters. The removal efficiencies for three predominant HAA species—Cl_2AA, Cl_3AA, and BrClAA—in both filters are summarized in Table 5.2.

From Table 5.2, it could be concluded that HAA removal between filter 1 (new) and filter 2 (6½ years old) was similar. Water temperature also affected the HAA removal.

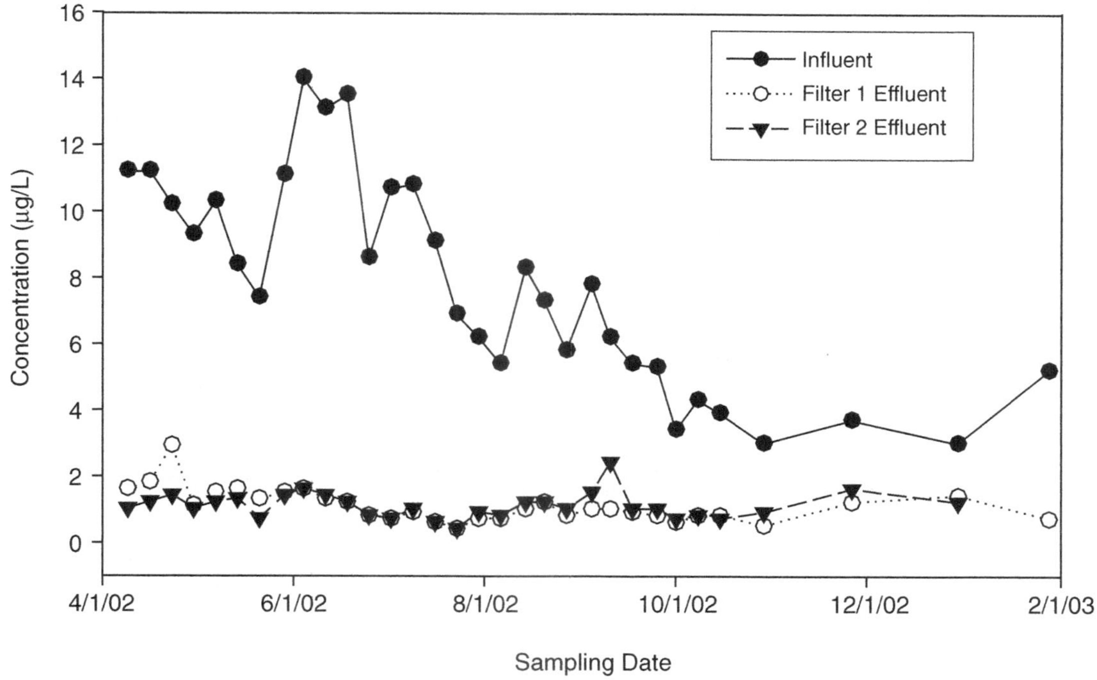

Figure 5.10 Cl_2AA influent and effluent concentrations for GAC filters in plant D

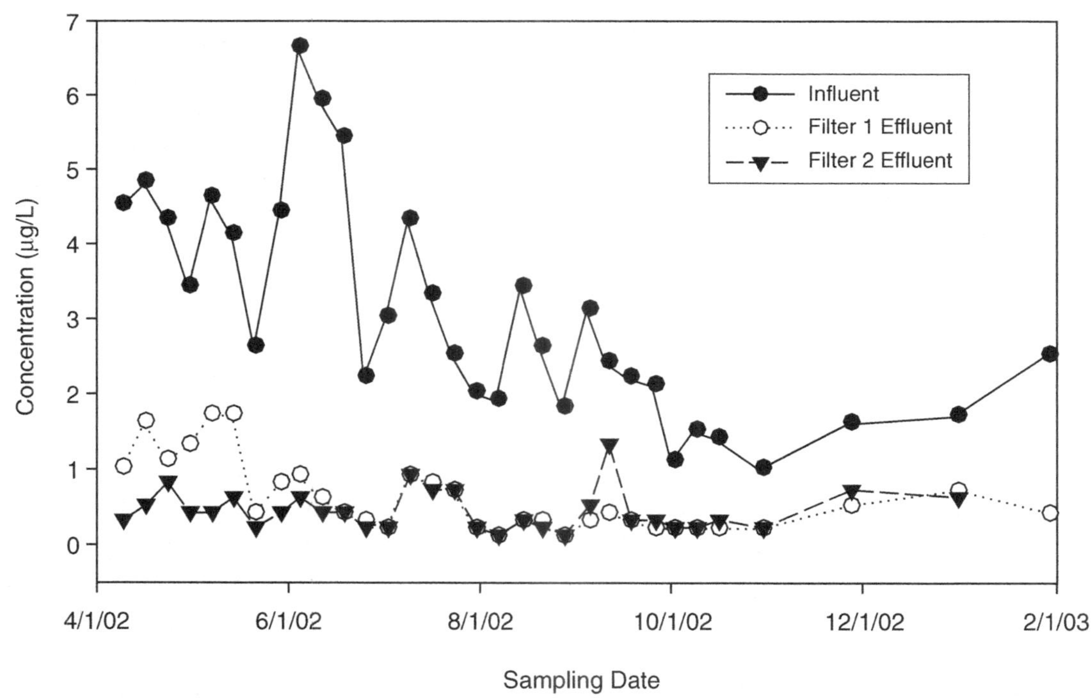

Figure 5.11 Cl_3AA influent and effluent concentrations for GAC filters in plant D

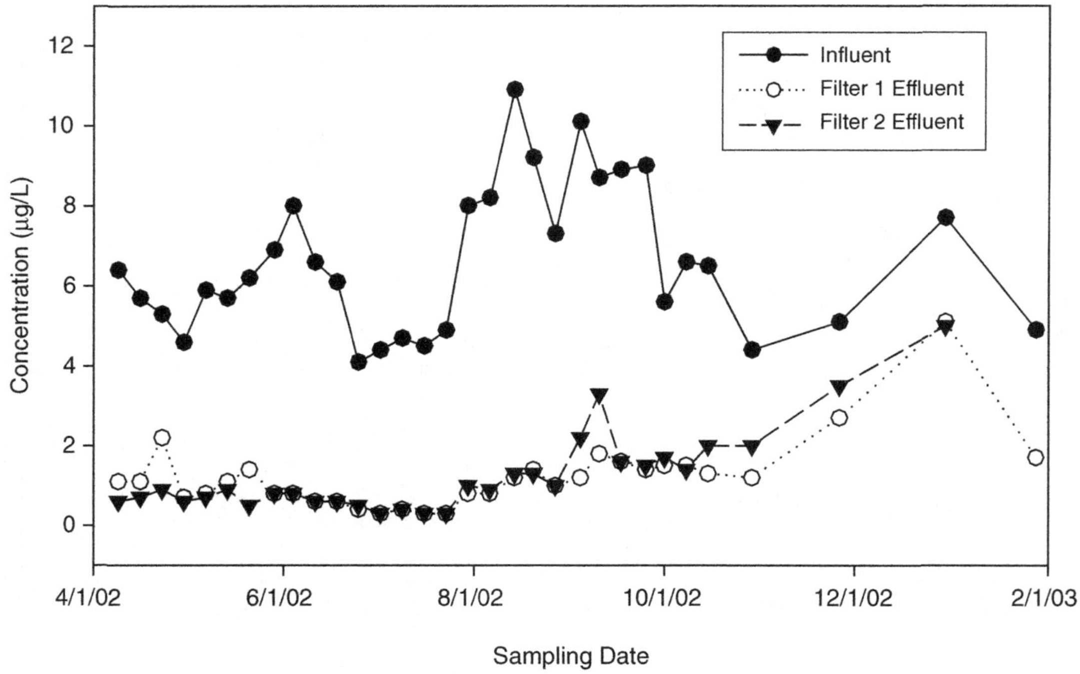

Figure 5.12 BrClAA influent and effluent concentrations for GAC filters in plant D

Table 5.2
Removal efficiencies for Cl_2AA, Cl_3AA, and BrClAA in two filters

	Cl_2AA		Cl_3AA		BrClAA	
Removal	Filter 1	Filter 2	Filter 1	Filter 2	Filter 1	Filter 2
Highest and date of removal	94%, July 23	94%, July 23	95%, Aug. 6	95%, Aug. 6	94%, July 23	94%, July 23
Lowest and date of removal	53%, Dec. 30	57%, Nov. 26	59%, Dec. 30	56%, Nov. 26	34%, Dec. 30	31%, Nov. 26
Average (%)	84	84	81	84	81	81

Even with a shorter EBCT, plant D had a Cl_2AA removal efficiency similar to plant A. The Cl_3AA removal efficiency in plant D was slightly higher than that in plant A. This may be due to the warmer climate associated with plant D (the San Francisco, Calif., area).

In plant D, in addition to $CHCl_3$, brominated THMs were also detected in the filter influent. Three sets of THM samples were collected after new GAC had been online for 7 months. No THM removal was observed in filter 1 (7 months old) and filter 2 (7 years old).

Table 5.3
Effects of GAC filtration on DBP levels in distribution systems (April)

	System 1 with GAC THMs:HAA5 (ratio)	System 2 with GAC THMs:HAA5 (ratio)	System 3 with GAC THMs:HAA5 (ratio)
GAC filter influent	11.7:26.0 (0.45)*	24.8:37.0 (0.67)*	70.5:93.3 (0.75)*
Finished water	23.5:21.5 (1.09)	37.4:30.2 (1.24)	102.1:47.2 (2.16)*
25% distribution	25.0:24.0 (1.04)	38.2:33.6 (1.14)	92.8:50.6 (1.83)
50% distribution	28.3:28.0 (1.01)	41.7:36.9 (1.13)	88.6:39.7 (2.23)
Maximum distribution	29.4:28.0 (1.07)	88.4:54.8 (1.61)†	88.4:54.8 (1.61)†
	System 4 without GAC	**System 5 without GAC**	**System 6 without GAC**
Finished water	18.6:29.8 (0.62)	26.2:42.1 (0.62)	21.7:47.0 (0.46)
25% distribution	25.2:40.9 (0.62)	29.3:44.3 (0.66)	45.3:88.7 (0.51)
50% distribution	28.2:41.2 (0.68)	37.3:65.6 (0.57)	40.0:78.6 (0.51)
Maximum distribution	35.6:38.6 (0.92)	35.2:60.1 (0.59)	56.8:84.1 (0.68)

*Average DBP values for two trains or plants.
†Distribution system shared by two systems.

Effects of GAC Filtration on HAA Levels in the Distribution System

GAC filtration could significantly reduce HAA5 levels in the distribution system. Because of the continuous formation of DBPs, including both THMs and HAAs, in distribution systems and negligible removal of THMs in aged GAC filters (more than 3 months old), the ratio of THMs to HAA5 was used to illustrate the effect of GAC filtration on HAA removal.

The April sampling results indicated that the average ratio (0.62) in GAC influents for three GAC systems was similar to that (0.57) in finished water for three non-GAC systems, as shown in Table 5.3. GAC filtration significantly increased the average ratio from 0.62 in the GAC influents to 1.50 in GAC effluents and 1.41 in the distribution systems of three GAC water systems. The decrease of the ratio from 1.50 in GAC effluents to 1.41 in the distribution systems indicates the continuous formation of HAAs in finished water. This average ratio (1.41) in distribution systems for three GAC systems was also much higher than the ratio (0.64) obtained in the distribution systems for three non-GAC systems. This indicates selective removal of HAA5 by GAC filtration even though HAAs were continuously formed in finished water.

The August sampling results also indicate that GAC filtration significantly increased the ratio from 1.10 in the GAC influents to 4.60 in GAC effluents and 3.10 in the distribution systems of three GAC water systems, as shown in Table 5.4. For three non-GAC systems, the ratio (1.23) in distribution systems was higher than that (0.94) in finished water. This increase indicated biological degradation of HAAs in distribution systems, as reported by Williams, Rindfleisch, and Williams (1995) and Williams, Williams, and Gordon (1996). Biological degradation of HAAs was also observed in the April sampling (from 0.57 to 0.64) but was less notable because of the low water temperature.

Table 5.4
Effects of GAC filtration on DBP levels in distribution systems (August)

	System 1 with GAC THMs:HAA5 (ratio)	System 2 with GAC THMs:HAA5 (ratio)	System 3 with GAC THMs:HAA5 (ratio)
GAC filter influent	34.7:36.9 (0.94)	41.2:36.6 (1.12)*	111.1:89.5 (1.24)*
Finished water	54.4:16.7 (3.26)	58.3:19.6 (2.97)	120.2:15.9 (7.56)*
25% distribution	53.5:21.9 (2.44)	47.9:22.5 (2.13)	138.8:30.8 (4.51)
50% distribution	55.1:21.2 (2.60)	66.0:26.3 (2.51)	110.2:22.7 (4.85)
Maximum distribution	72.5:32.1 (2.26)	123.4:37.6 (3.28)†	123.4:37.6 (3.28)†
	System 4 without GAC	System 5 without GAC	System 6 without GAC
Finished water	26.9:30.5 (0.88)	36.6:27.9 (1.31)	31.8:50.3 (0.63)
25% distribution	51.3:55.5 (0.92)	49.3:38.8 (1.27)	91.5:56.2 (1.63)
50% distribution	66.0:59.4 (1.11)	76.5:75.8 (1.01)	103.3:135.5 (0.76)
Maximum distribution	47.3:50.3 (0.94)	93.4:75.0 (1.25)	149.4:67.8 (2.20)

*Average DBP values for two trains or plants.
†Distribution system shared by two systems.

Under the Stage 1 D/DBP Rule, THMs and HAA5 are regulated with MCLs of 80 μg/L and 60 μg/L, respectively (USEPA 1998). For two non-GAC systems—5 and 6—HAA5 levels in five of the six samples were higher than the MCL (60 μg/L), even though THM levels in all six samples were lower than the MCL (80 μg/L). The relative dominance of HAAs in these water samples is comparable to that reported in North Carolina water systems by Singer, Obolenski, and Greiner (1995). This implies that HAAs are more likely to trigger the MCL violation under the Stage 1 D/DBP Rule in North Carolina, Pennsylvania, and other states with water low in bromide. However, in two GAC systems—2 and 3—GAC filtration reduced the HAA5 levels below the MCL (60 μg/L) in all six samples even though the THM levels were higher than the MCL (>80 μg/L) in four of the six samples.

The August sampling results gave a similar trend. For two non-GAC systems—5 and 6—HAA5 levels in four of the six samples were higher than the MCL (60 μg/L). In comparison to the April samples, the relatively low HAA levels in three non-GAC systems may be due to biological degradation of HAAs in distribution systems, especially in system 6. However, in two GAC systems—2 and 3—GAC filtration reduced the HAA5 levels below the MCL (60 μg/L) in all six samples even though the THM levels were higher than the MCL (>80 μg/L) in four of the six samples. In system 3, GAC filtration reduced HAA5 from 89.5 μg/L in the GAC influent to 15.9 μg/L in the GAC effluent and 30.4 μg/L (average) in the distribution system.

HAA biodegradation in distribution systems may play a role in reducing HAA5 levels in these systems. However, biodegradation alone may not be capable of lowering HAA5 below the MCL as indicated by two non-GAC systems—5 and 6.

The distribution system study indicates that GAC filtration is an effective technology to control HAAs in treated water and distribution systems. Even though HAAs and other DBPs are continuously formed in the systems, GAC filtration is a viable technology for HAA compliance with the D/DBP Rule.

CONCLUSIONS

The results indicate that GAC filters effectively reduced HAA concentrations in their effluents. For HAA removal, there were no significant differences between new and aged GAC filters, especially after a few months of operation. The findings also show that EBCT affected the HAA removal, especially for Cl_3AA, and that GAC filters with a longer EBCT had a better removal of HAAs. Water temperature also affected HAA removal, especially for Cl_3AA. In winter, no Cl_3AA removal was observed in GAC filters, especially in those with a short EBCT.

GAC filtration could significantly reduce HAA5 levels in distribution systems. In three GAC water systems, GAC filtration significantly increased the ratios of THMs to HAA5. For two non-GAC systems—5 and 6—HAA5 levels in five of the six samples were higher than the MCL (60 μg/L). However, in two GAC systems—2 and 3—GAC filtration reduced the HAA5 levels below the MCL (60 μg/L) in all six samples even though the THM levels were higher than the MCL (>80 μg/L) in four of the six samples. In spite of the fact that HAAs and other DBPs are continuously formed in the distribution system, GAC filtration would be a viable technology for HAA5 compliance.

CHAPTER 6
TECHNIQUES TO ACCELERATE BIOACTIVITY DEVELOPMENT

INTRODUCTION

Several studies have reported the removal of HAAs in BAC filters through biodegradation (Xie and Zhou 2002, Singer et al. 1999, Wobma et al. 2000). As reported by Xie and Zhou (2002), the removal of HAAs in a new GAC filter has three distinctive phases: physical adsorption, carbon saturation, and biological degradation. Because of the low GAC adsorption capacity for HAAs (Tung, Unz, and Xie 2003), carbon saturation for HAAs generally occurs in 1 or 2 months. A higher level of HAAs is then expected in the effluent, because biological activity development generally lags behind carbon saturation. Therefore, accelerating bioactivity development in GAC filters could expedite the biological degradation phase and reduce HAAs in the effluent. Many factors could affect bioactivity development in a GAC filter. Seeding is a common way to accelerate biological growth in a biological reactor. Backwashing with nonchlorinated water also promotes bioactivity development in biofilters (Reckhow et al. 1992).

In this task, several techniques were explored for accelerating the bioactivity development on GAC. These included seeding the new GAC filter with acclimated GAC or acclimated sand medium, and quenching chlorine residual and adding nitrate for HAA removal in new GAC filters. The effects of filter backwash were evaluated as well.

MATERIALS AND METHODS

Synthetic Water Samples

Pilot GAC filters were set up and run in a local water treatment plant. The plant consists of conventional water treatment processes, including coagulation, sedimentation, and filtration. The plant effluent, with a free chlorine residual of approximately 2.5 mg/L and a nitrate concentration of 0.1 mg/L, was spiked with HAA5 at 50 μg/L each and used as the pilot GAC filter influent. HAA concentrations in influent and effluent samples were analyzed using modified USEPA Method 552.2 (USEPA 1995a; Xie, Reckhow, and Springborg 1998).

Effects of Seeding

Three 5.1-cm (2-in.) columns were packed, to a depth of 76.2 cm (30 in.), with virgin GAC (F-100) obtained from Calgon Carbon. Column 1 was packed with 100% virgin GAC, column 2 with 95% virgin GAC and 5% acclimated GAC, and column 3 with 100% virgin GAC on the top and 15.2 cm (6 in.) of acclimated sand at the bottom. Both acclimated GAC and acclimated sand samples were collected from GAC–sand filters at local water treatment plants. The three columns were backwashed once every other day. Influent and effluent water samples were collected daily. The first run was started at a high water temperature (20°C), and the second run was started at a low water temperature (11°C).

Effects of Filter Backwash

Two 5.1-cm (2-in.) columns were packed with 76.2 cm (30 in.) of acclimated GAC each and run continuously for 2 days. Then the two GAC columns were backwashed using different water flow rates for 15 minutes. The filter backwash water flow rates were set by the filter bed expansion, which were 100% for column 1 and 30% for column 2.

Influent and effluent samples were taken three times with a time interval of 30 minutes before the backwash. After backwashing the columns, influent and effluent water samples were collected at 0, 10, 40, 70, 100, and 160 minutes, and at 24 and 48 hours.

Effects of Chlorine Residual Quenching and Nitrate Addition

Three 5.1-cm (2-in.) columns were packed with 76.2 cm (30 in.) of virgin GAC (F-400). The filters were operated at an EBCT of 5 minutes. The synthetic water sample, or finished water spiked with HAA5, was used as the influent for column 1. The second aliquot of the synthetic water sample was spiked with 5 mg/L sodium nitrate and used as the influent for column 2. The third aliquot of the synthetic water sample was added with 10 mg/L sodium thiosulfate to quench chlorine residual and used as the influent for column 3. Influent and effluent water samples were collected once every other day right before the backwash.

RESULTS AND DISCUSSION

Effects of Seeding

Under the first run, column 1 and column 2 were started on October 1, 2002. The water temperature was around 20°C for the first 10 days and then decreased gradually, as shown in Figure 6.1. HAA influent and effluent concentrations for column 1 and column 2 are shown in Figures 6.2–6.6.

Without seeding, GAC in column 1 reached the saturation for ClAA on day 4. No ClAA was removed between day 4 and day 10. After day 11, ClAA concentration in the effluent decreased gradually. The effluent ClAA concentration was nearly stable after day 25. The average ClAA concentration in the effluent was 20.4 μg/L, and the average removal efficiency was 62.2%. This pattern of the ClAA effluent concentration is very similar to that reported by Xie and Zhou (2002) and indicated that the adsorption capacity of the GAC filter was completely exhausted after day 4. After day 11, biological activities were established and biological degradation of ClAA started. After day 25, biological activities in the GAC filter were fully established.

In column 2, seeded with acclimated GAC, the concentration of ClAA was always lower than that in column 1. The additional removal in column 2 indicated biological degradation of ClAA. Because of accelerated biological activity development, the saturation period (after carbon exhaustion and before the biodegradation) never appeared. Figure 6.2 also indicates that biological activity was fully established in column 2 after day 16, which is much shorter than in column 1 without seeding. The stable removal efficiency for column 2 was 72%.

The removal patterns of BrAA, Cl_2AA, and Br_2AA in the two columns were similar to that of ClAA, as shown in Figures 6.3, 6.4., and 6.5. The results indicate that seeding with 5% acclimated GAC enhanced HAA removal at the early stage of GAC filter operation. Seeding with 5% acclimated GAC also shortened the time required for biological activity to be fully developed.

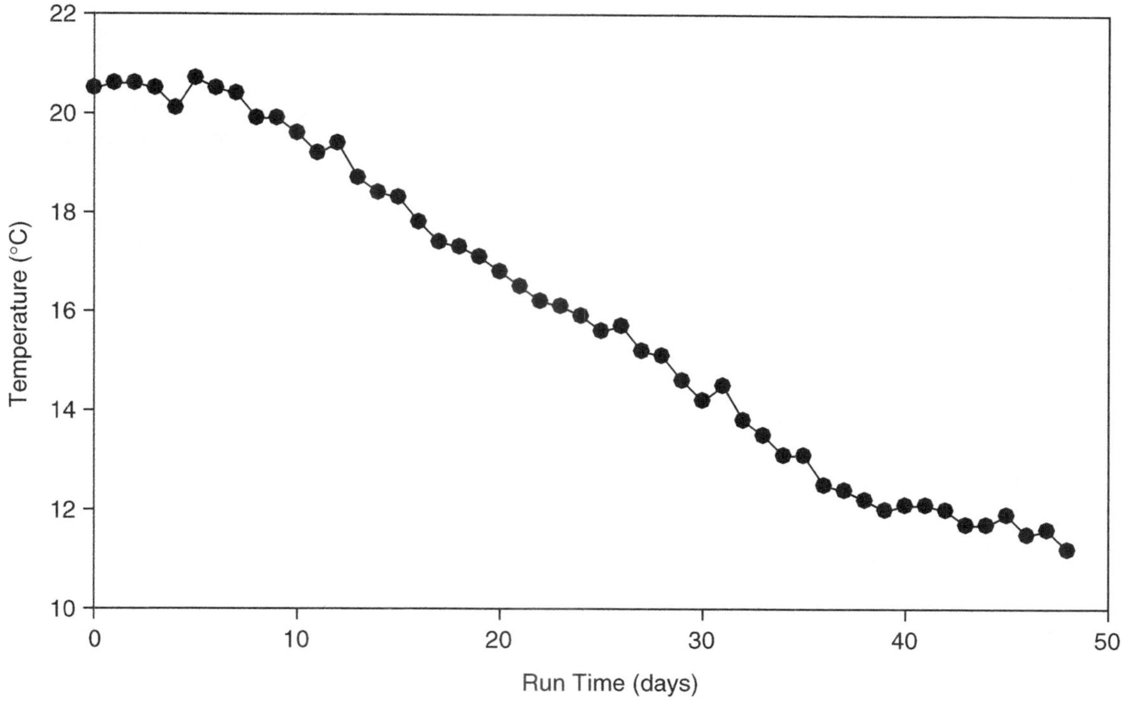

Figure 6.1 Influent water temperatures (the first run)

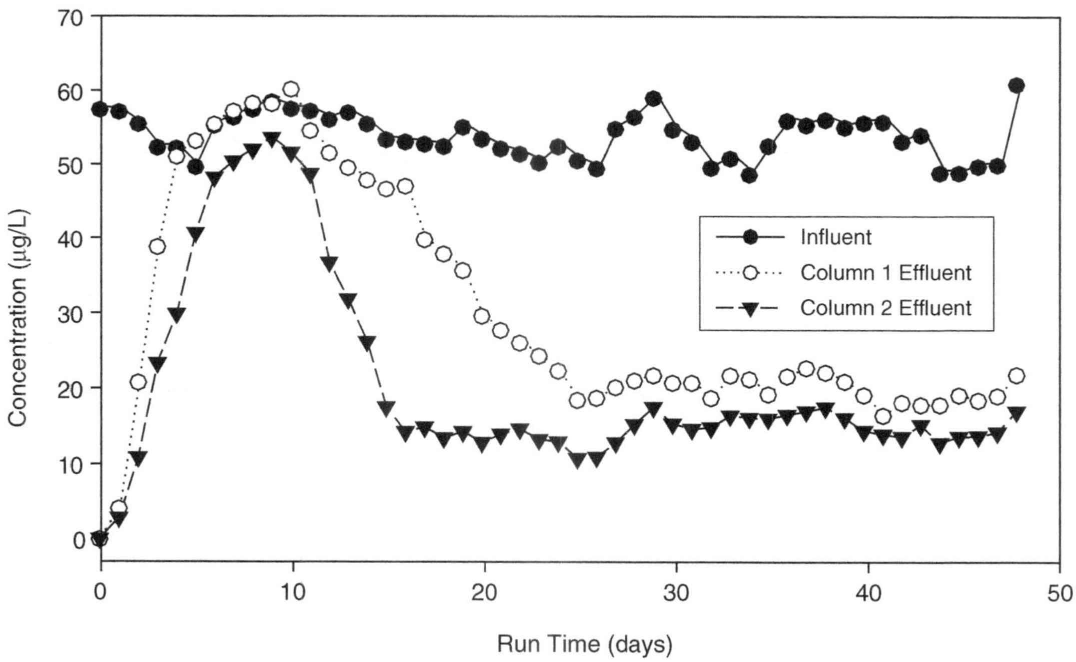

Figure 6.2 Effects of seeding on ClAA removal

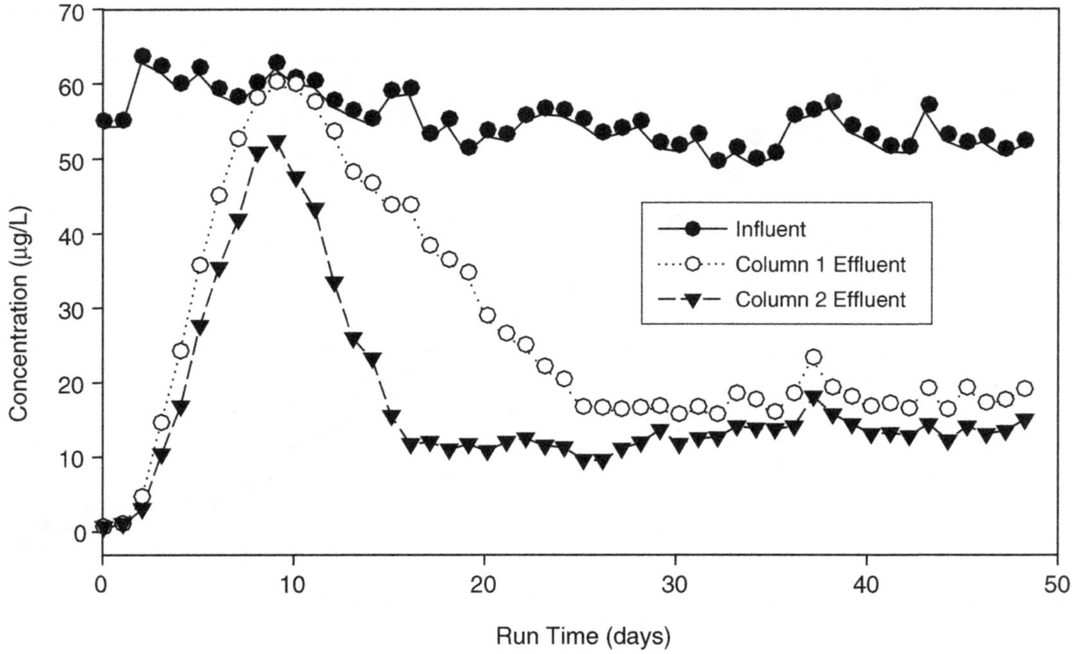

Figure 6.3 Effects of seeding on BrAA removal

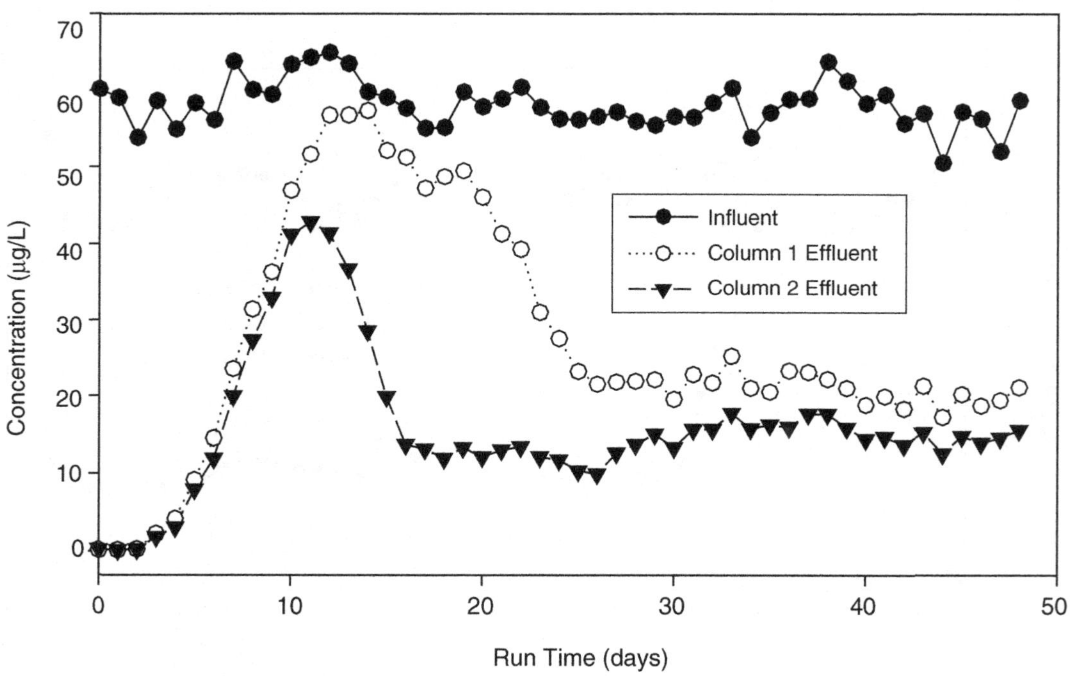

Source: Xie, Wu, and Tung 2003.

Figure 6.4 Effects of seeding on Cl_2AA removal

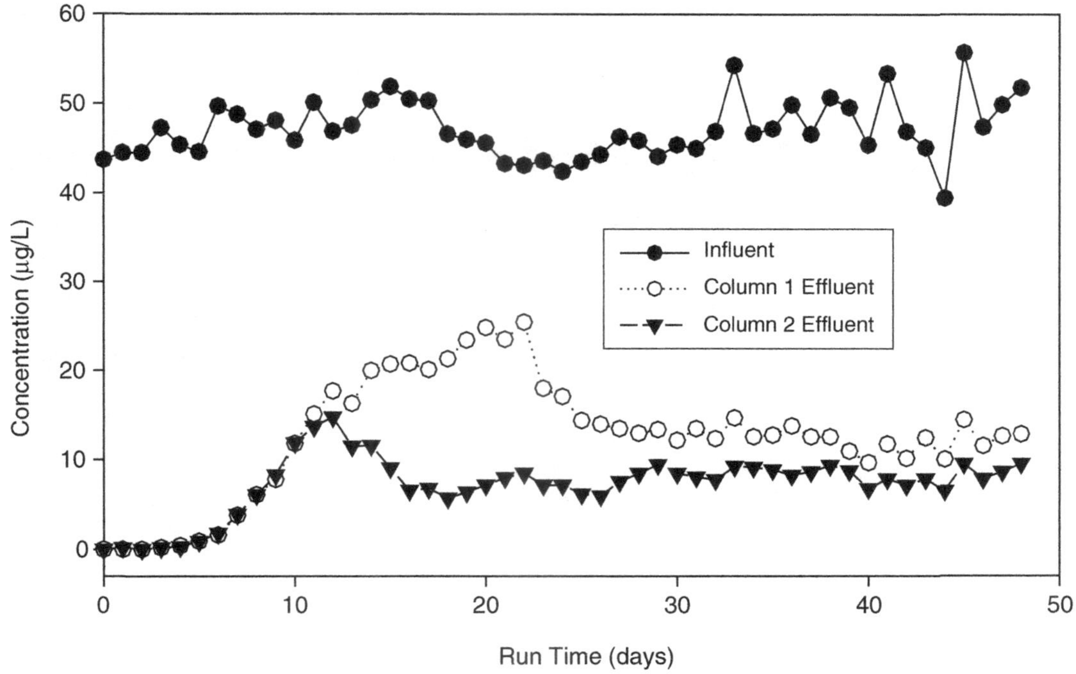

Figure 6.5 Effects of seeding on Br_2AA removal

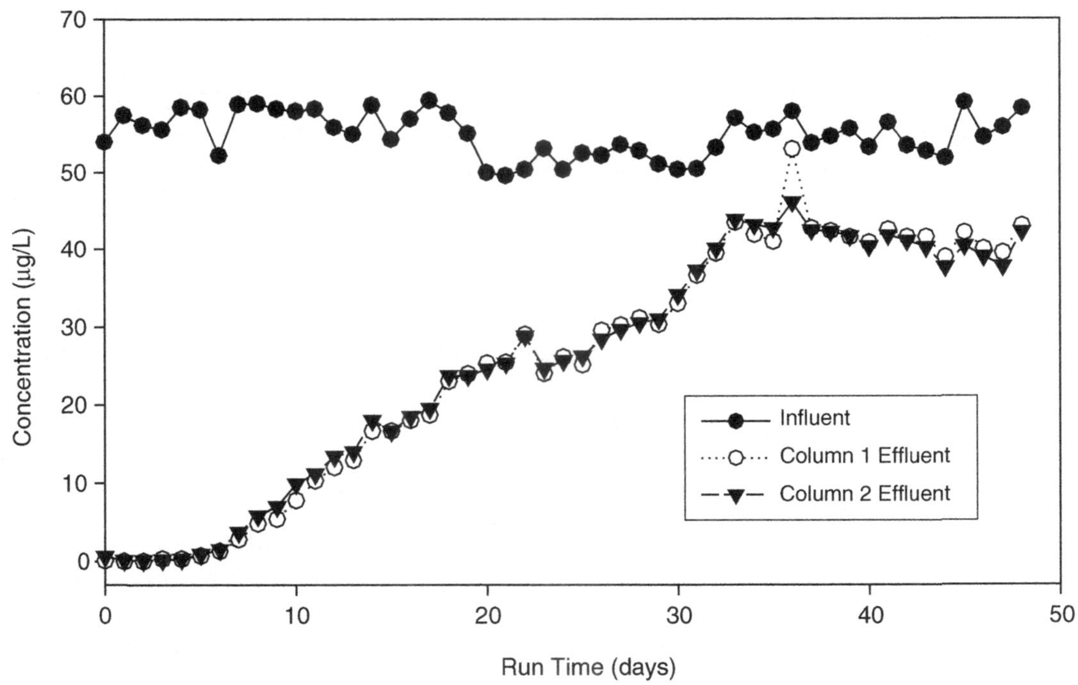

Source: Xie, Wu, and Tung 2003.

Figure 6.6 Effects of seeding on Cl_3AA removal

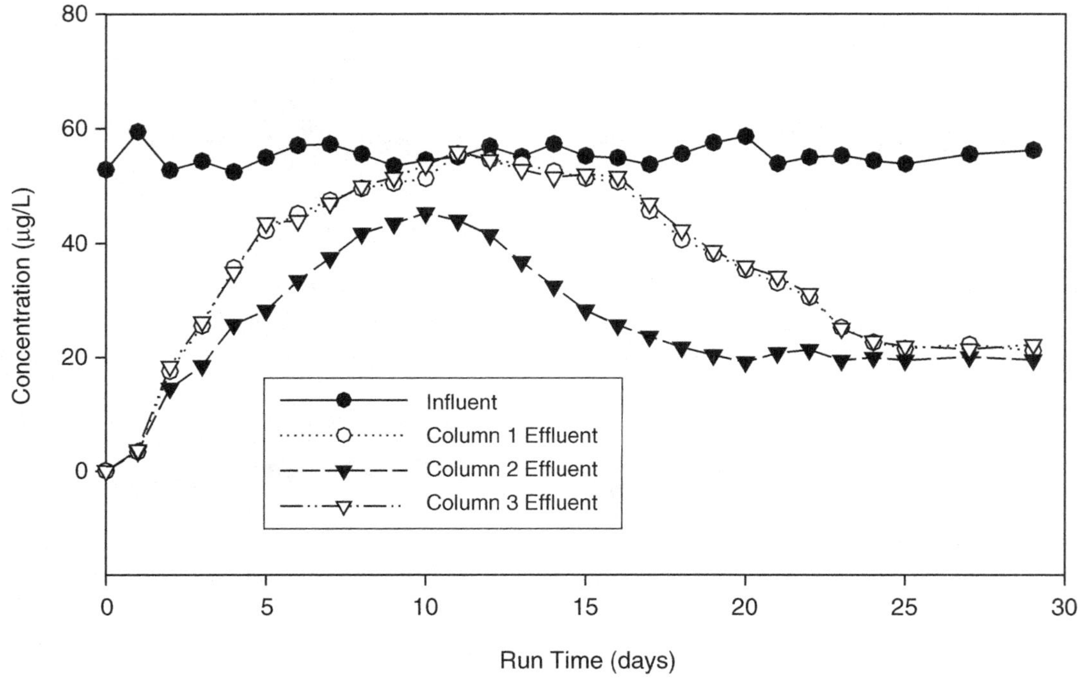

Figure 6.7 Effects of seeding on ClAA removal

The Cl_3AA effluent concentrations for column 1 and column 2 are shown in Figure 6.6. There was little difference between Cl_3AA effluent concentrations in columns 1 and 2. Increasing the run time raised Cl_3AA effluent concentration in both columns. This indicates that GAC did not reach saturation for Cl_3AA within the 50-day run time and suggests a high GAC adsorption capacity for Cl_3AA, as reported by Tung, Unz, and Xie (2002). The identical Cl_3AA removal efficiency for columns 1 and 2 indicates that seeding with acclimated GAC had no impact on Cl_3AA removal. This is consistent with the fact that Cl_3AA is not readily biodegradable, as reported by Tung, Unz, and Xie (2003).

Under the second run, three GAC columns were started on February 19, 2003, at a water temperature of 11°C. During the run, the water temperature increased gradually. In comparison to the first run, similar HAA removal results under the second run were obtained for column 1, as shown in Figures 6.7–6.11. Column 3, seeded with acclimated sand, and column 1 had almost identical HAA removal performances, indicating that seeding with acclimated sand did not affect the bioactivity development in column 3. This may be due to little biological activity on the acclimated sand. Similar to the results obtained at a high temperature, seeding with acclimated GAC reduced the maximum HAA concentration in the effluent and the time required for bioactivity development. The steady effluent concentration of column 2 was also lower than those of columns 1 and 3. Therefore, column 2 had better removal for HAA than columns 1 and 3. Because of the water temperature variation during two runs, the effects of water temperature on the biological activity development in GAC filters were not assessed.

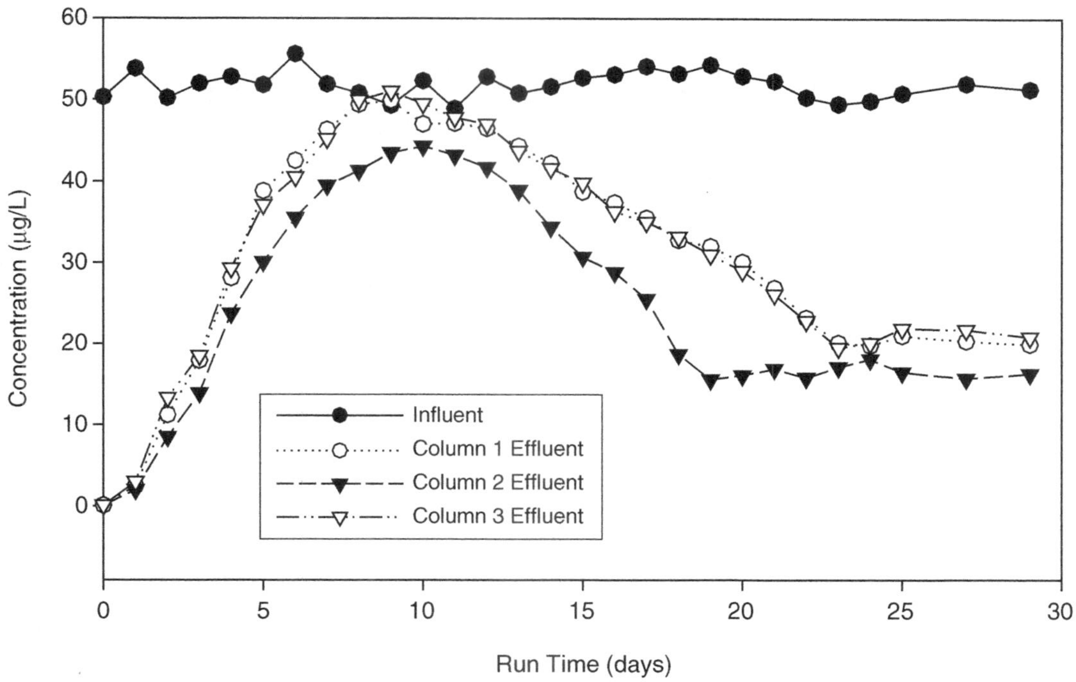

Figure 6.8 Effects of seeding on BrAA removal

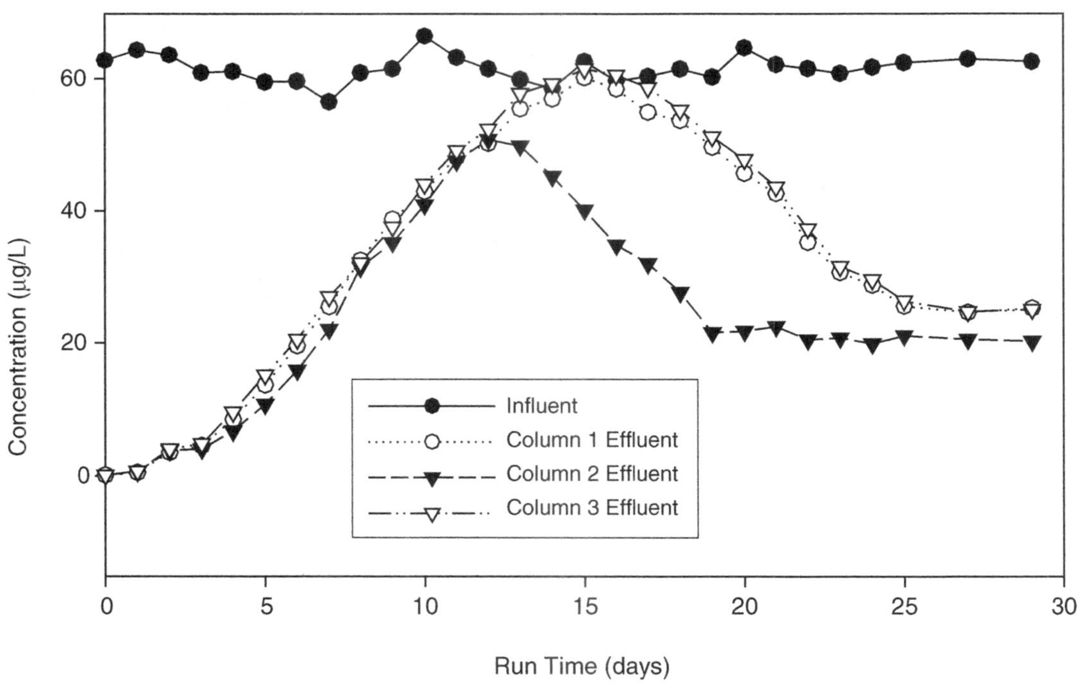

Figure 6.9 Effects of seeding on Cl_2AA removal

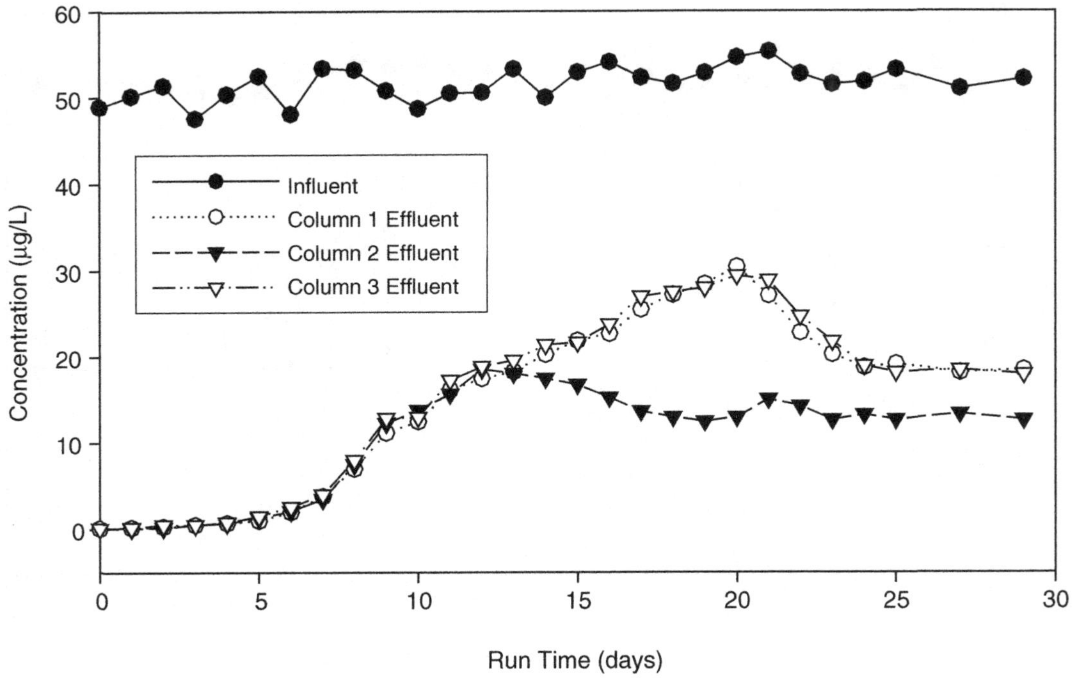

Figure 6.10 Effects of seeding on Br_2AA removal

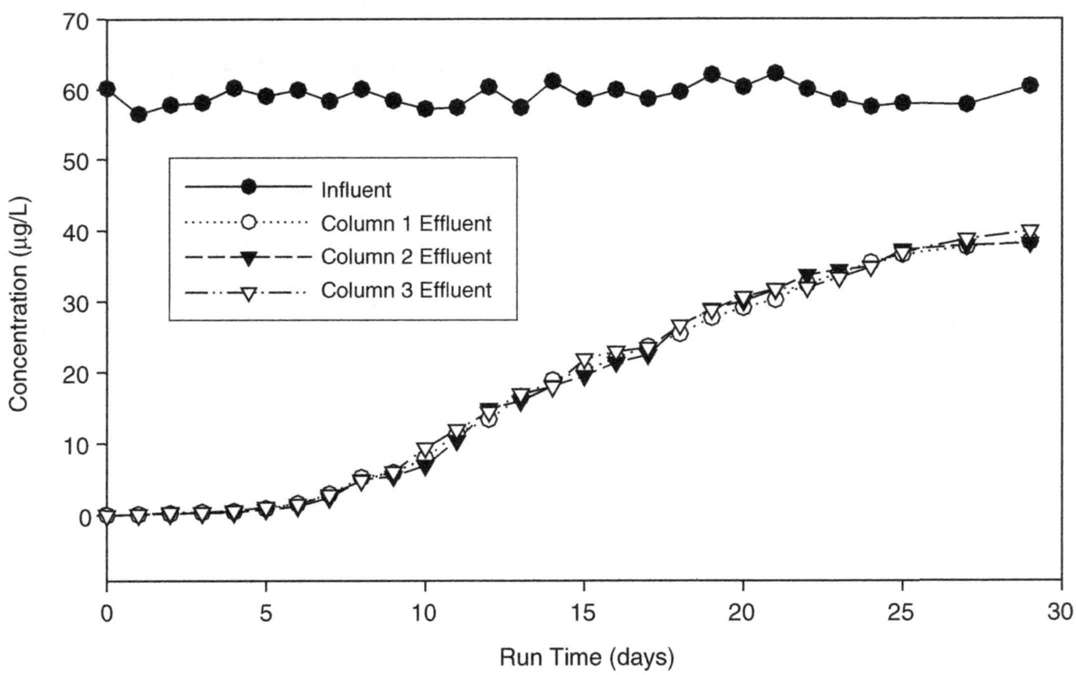

Figure 6.11 Effects of seeding on Cl_3AA removal

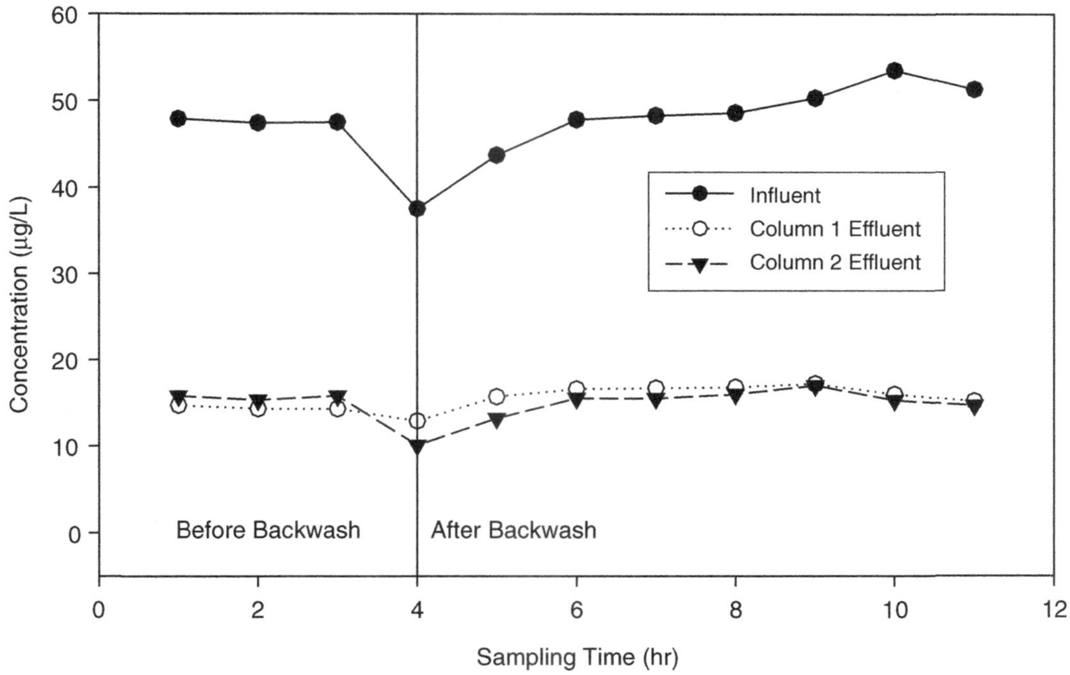

Figure 6.12 Effects of backwash on ClAA removal

Effects of Filter Backwash

Effects of backwash on HAA removal in BAC filters are shown in Figures 6.12–6.14. Column 1 was backwashed at a flow rate of 77.0 m/hr (31.5 gpm/ft^2; 100% bed expansion), and column 2 was backwashed at 35.4 m/hr (14.5 gpm/ft^2; 30% bed expansion). The backwash water contained approximately 2.5 mg/L free residual chlorine.

Backwash with chlorinated effluent had little effect on ClAA removal. After backwash, a slightly better removal of ClAA was obtained in column 2, which had a lower backwash rate. The difference between ClAA removal at both backwash rates became negligible when the filter run time increased. The reduction in the ClAA influent concentration right after the filter backwash was due to the dilution of backwash water, which contained lower level of HAAs. Similar results were observed for Cl_2AA (Figure 6.13) and Cl_3AA (Figure 6.14). Filter backwashing with chlorinated water at both backwash rates did not significantly affect HAA removal.

Effects of Chlorine Residual Quenching and Nitrate Addition

Quenching residual chlorine in the filter influent (column 3) had little impact on HAA removal. In comparison to no quenching (column 1), quenching chlorine residual resulted in an early ClAA breakthrough in column 3, as shown in Figure 6.15. This may be due to competition for adsorption sites between thiosulfite and ClAA. Similar effects were also observed for Cl_2AA and Cl_3AA, as shown in Figures 6.16 and 6.17.

For ClAA and Cl_2AA, column 3 took longer for its effluent level to come down than column 1, as shown in Figures 6.15 and 6.16. This indicates that quenching residual chlorine in the filter influent also delayed the biological activity development. Lower HAA concentrations on

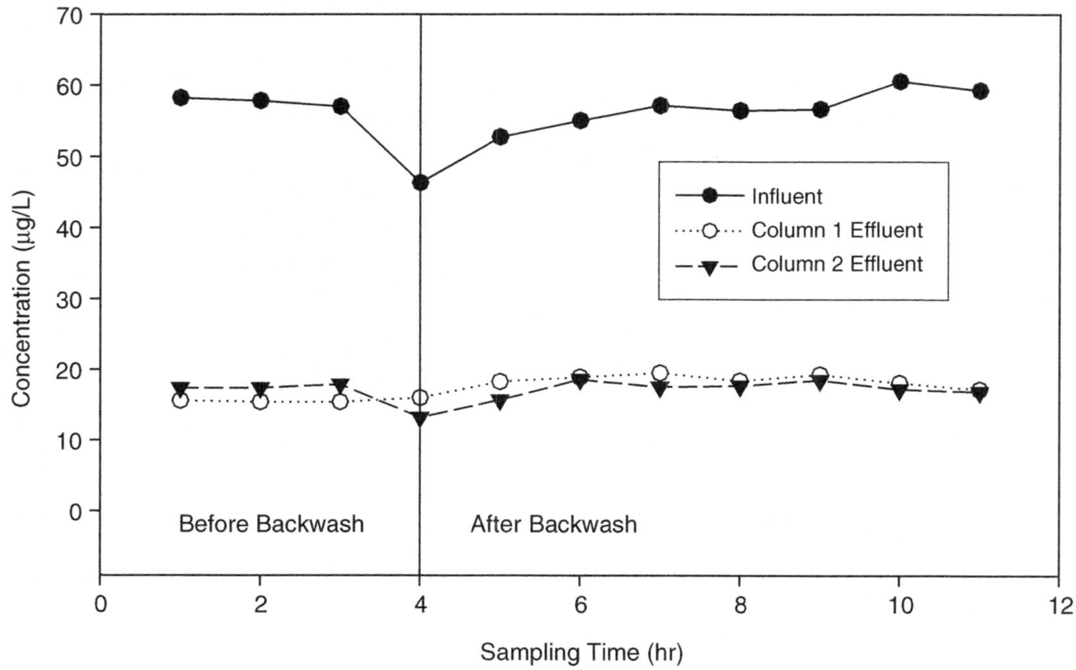

Figure 6.13 Effects of backwash on Cl_2AA removal

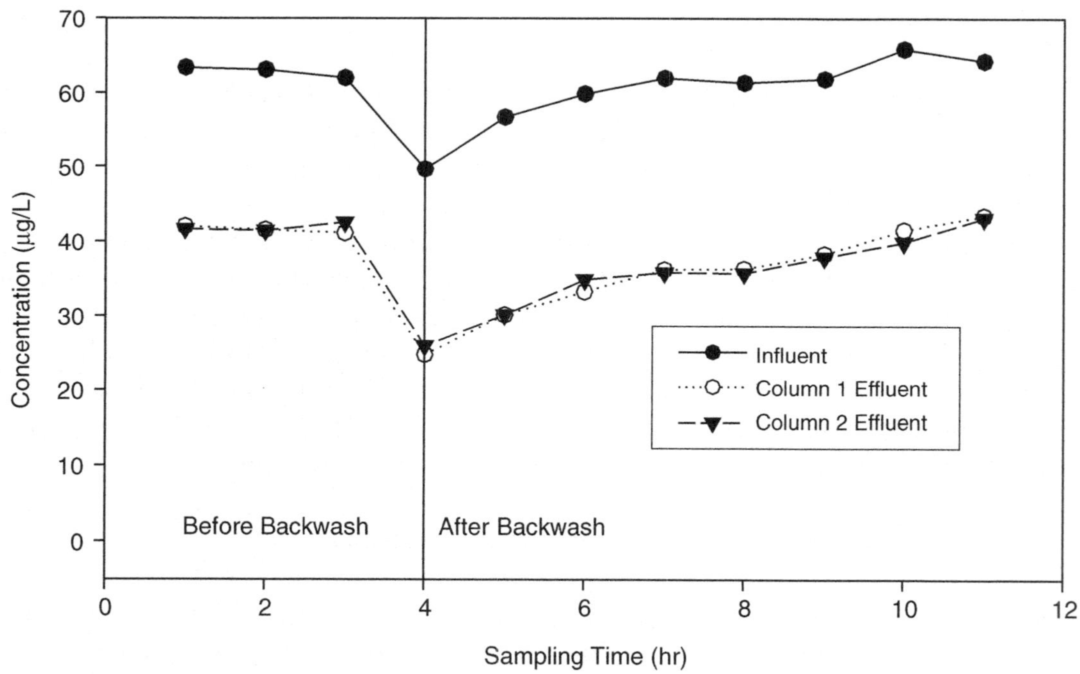

Figure 6.14 Effects of backwash on Cl_3AA removal

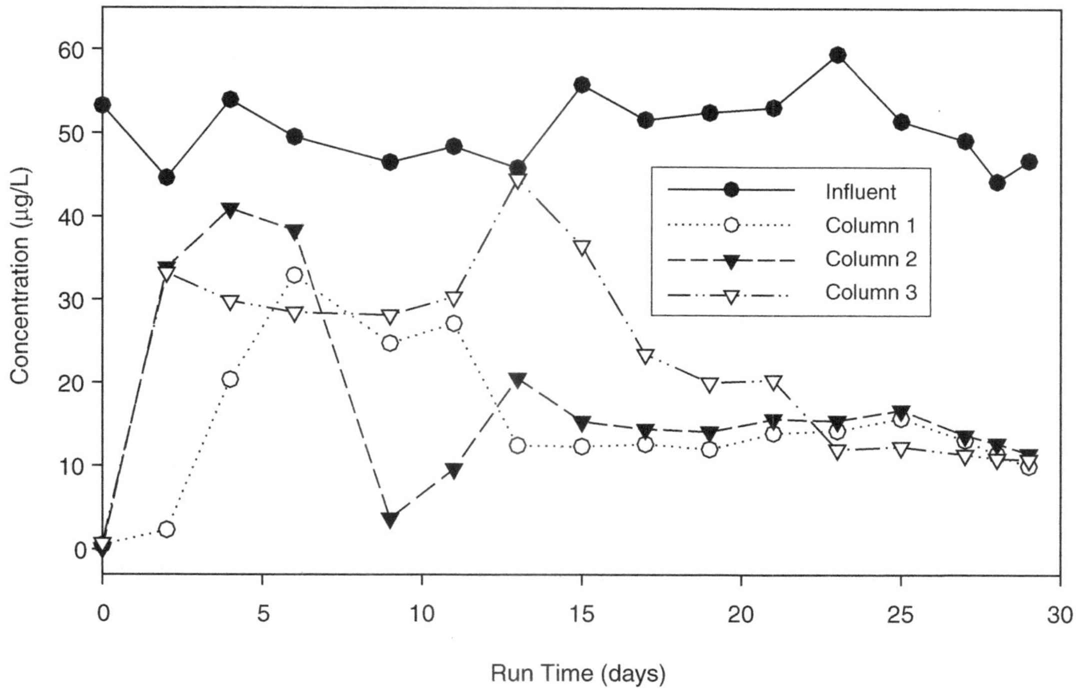

Figure 6.15 Effects of chlorine quenching and nitrate on ClAA removal

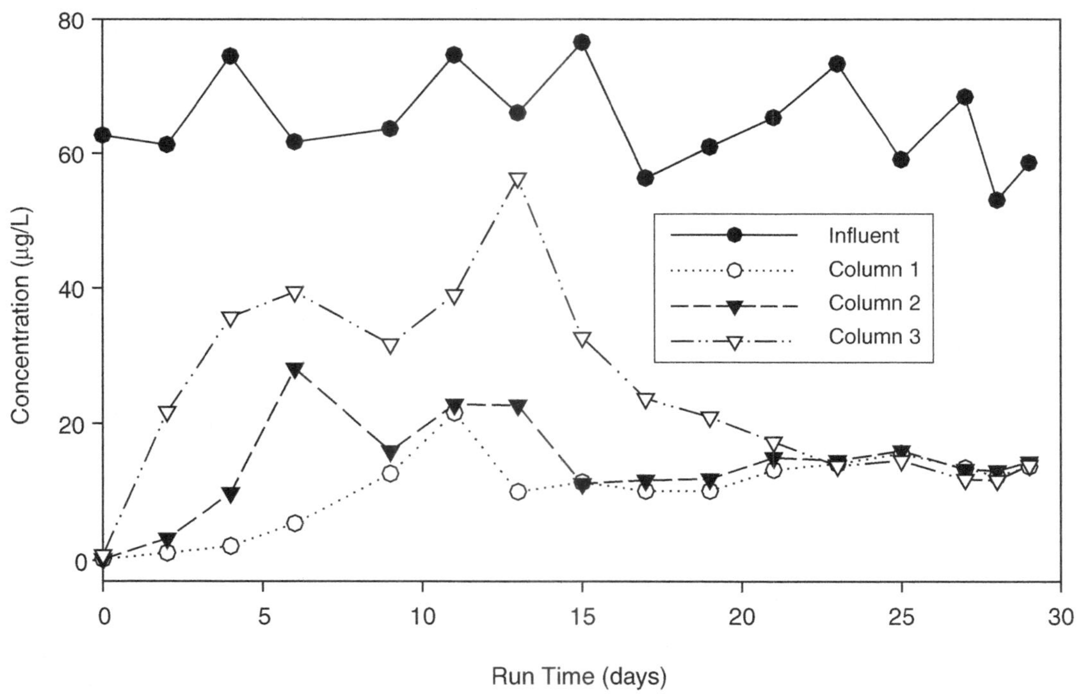

Figure 6.16 Effects of chlorine quenching and nitrate on Cl_2AA removal

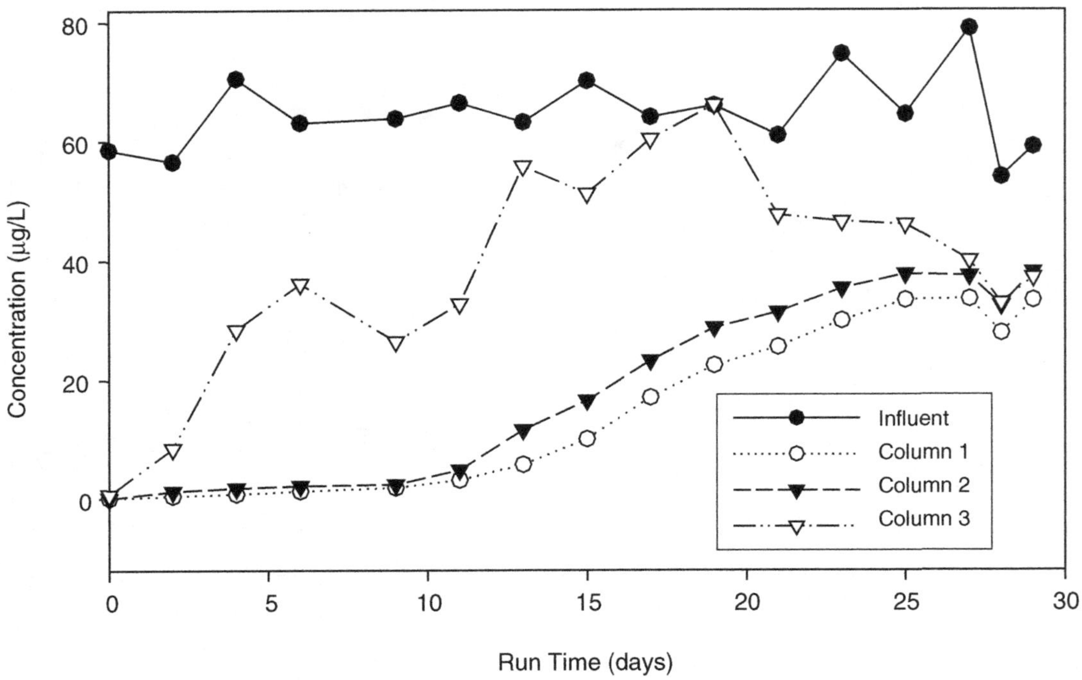

Figure 6.17 Effects of chlorine quenching and nitrate on Cl_3AA removal

the GAC surface, due to thiosulfate competition, may delay the biological activity development on GAC. For Cl_3AA, biodegradation was not observed in either column 1 or column 3.

Adding nitrate in column 2 resulted in an early breakthrough for ClAA, Cl_2AA, and Cl_3AA, as shown in Figures 6.15, 6.16, and 6.17. Again, this may be due to competition for adsorption sites between nitrate and HAAs. The degree of impact from nitrate was less than that from thiosulfite, due perhaps to their charge difference (–1 versus –2) and concentration difference (5 mg/L versus 10 mg/L). After the saturation phase, there was little difference in HAA effluent concentrations between column 1 and column 2. In this preliminary study, adding nitrate had no noticeable effect on the biological activity development and biological degradation of HAAs.

No noticeable effects of chlorine residual quenching and nitrate addition were obtained in this preliminary study except for their competing effect. Further study is needed to closely examine the effect of chlorine residual and nitrate on biological activity development in a new GAC filter and biological degradation of HAAs in a BAC filter.

CONCLUSIONS

The results indicate that seeding with 5% acclimated GAC enhanced HAA removal (except Cl_3AA) at the early stage of a GAC filter operation. The seeding also shortened the time required for biological activity to be fully developed. However, seeding with acclimated GAC had no impact on Cl_3AA removal. Seeding with acclimated sand did not affect the bioactivity development in a GAC filter.

Backwash with chlorinated water at both backwash rates did not significantly affect HAA removal in BAC filters. Quenching chlorine residual with sodium thiosulfite resulted in an early

HAA breakthrough and delayed the biological activity development. Lower HAA concentrations on the GAC surface, caused by thiosulfate competition, may delay the biological activity development on GAC. Adding nitrate had no noticeable effect on the biological activity development and biological degradation of HAAs except for an early HAA breakthrough during the adsorption phase of the GAC filter. The early HAA breakthrough may be due to the competition for GAC adsorption site between HAAs and thiosulfate or nitrate.

CHAPTER 7
CONCLUSIONS

TASK 1

The adsorption isotherm tests were conducted for six HAAs and four THMs. Their isotherm constants were calculated using the Freundlich equation. HAA species with a higher halogen number had a higher GAC adsorption capacity (K). For both HAAs and THMs, their brominated species had higher GAC adsorption capacities (K) than did their chlorinated analogues.

The isotherm constants indicate that the GAC adsorption capacities for THMs were higher than that for HAAs except for Cl_3AA. The full-scale GAC filter study also indicates that the GAC bed life for $CHCl_3$ is longer than that for ClAA, BrAA, Cl_2AA, BrClAA, and Br_2AA. This is contrary to what other studies reported and may be due to HAA biodegradation during adsorption isotherm tests conducted by others. Biodegradation of HAAs during isotherm tests will result in an overestimation of their adsorption capacity. The isotherm tests conducted in the presence of a minimal biological activity with the actual filter influent will improve the accuracy of GAC bed life estimation for HAA removal by adsorption.

The adsorption isotherm tests and the full-scale filter study indicate that the physical adsorption only plays a minor role in DBP removal in GAC filters.

TASK 2

The biofilm activity inhibition factors, including thermal stress (4°C and 50°C) and chloramine shock, significantly reduced the HAA removal rates in BAC columns. For the respiration characteristics of the HAA-degrading bacteria, an anoxic environment did not affect the HAA removal in column tests. In the batch test, all HAAs except Cl_3AA degraded under both aerobic and anoxic conditions. For HAA degradation, however, a lag phase was observed in anoxic conditions. Chloride release from HAAs incubated with BAC for 80 hours indicated complete dehalogenation for ClAA, 60% dehalogenation for Cl_2AA, and 15% dehalogenation for Cl_3AA.

These results indicate that biological dehalogenation, under both aerobic and anoxic conditions, plays a major role in HAA removal in BAC filters.

TASK 3

Results of this study indicate that both EBCT and water temperature significantly affected HAA removal in BAC columns. Raising the water temperature and EBCT increased the removal of HAAs. At a temperature of 20°C or higher, increasing EBCT had little effect on HAA removal. A longer EBCT could be used to compensate for the effect of a low water temperature. For an HAA6 removal efficiency of 50% or higher, the authors suggest a 10-minute EBCT for 4°C waters and a 5-minute EBCT for waters at 10°C or higher.

HAA removal in BAC columns had different kinetic models at different temperatures. At 4°C, HAA removal tended to follow a zero-order reaction model. This indicates that HAA removal may be limited by carbon adsorption capacity or biological activities on its surface. At

10°C, the monohaloacetic acid and dihaloacetic acid removal tended to follow a first-order reaction model. This indicates that biological activities were fully developed at 10°C and HAA removal was limited by HAA concentrations. For Cl_3AA removal, a first-order reaction model was obtained at 20°C but not at 10°C.

TASK 4

The results indicate that GAC filters effectively reduced HAA concentrations in their effluents. For HAA removal, there were no significant differences between new and aged GAC filters, especially after a few months of operation. EBCT affected HAA removal, especially for Cl_3AA. GAC filters with a longer EBCT had a better removal of HAAs. Water temperature also affected HAA removal, especially for Cl_3AA. In winter, no Cl_3AA removal was observed in GAC filters, especially in those with a short EBCT.

GAC filtration could significantly reduce HAA5 levels in distribution systems. In three GAC water systems, GAC filtration significantly increased the ratios of THMs to HAA5. For two non-GAC systems—5 and 6—HAA5 levels in five of the six samples from distribution systems were higher than the MCL (60 µg/L). However, in two GAC systems—2 and 3—GAC filtration reduced the HAA5 levels below the MCL (60 µg/L) in all six samples from distribution systems even though the THM levels were higher than the MCL (>80 µg/L) in four of the six samples. Despite the fact that HAAs and other DBPs are continuously formed in the distribution system, GAC filtration would be a viable technology for HAA5 compliance.

TASK 5

The results indicate that seeding with 5% acclimated GAC enhanced HAA removal at the early stage of a GAC filter operation. Seeding with 5% acclimated GAC also shortened the time required for biological activity to be fully developed. However, seeding with acclimated GAC had no impact on Cl_3AA removal. Seeding with acclimated sand did not affect the bioactivity development in a GAC filter.

Backwash with chlorinated water at both backwash rates did not significantly affect HAA removal in BAC filters. Quenching chlorine residual with sodium thiosulfite resulted in an early HAA breakthrough and delayed the biological activity development. Lower HAA concentrations on the GAC surface, caused by thiosulfate competition, may delay the biological activity development on GAC. Adding nitrate had no noticeable effect on the biological activity development and biological degradation of HAAs except for an early HAA breakthrough during the adsorption phase of the GAC filter. The breakthrough may be due to the competition between HAAs and thiosulfate or nitrate for the GAC adsorption site.

RECOMMENDATIONS

GAC filtration is an effective process for removing HAAs formed during prechlorination or intermediate chlorination. GAC filtration could also be used by consecutive systems for HAA control in their distribution systems. GAC filtration can provide a long-term (several years') removal for HAAs through biological degradation and a short-term (several months') removal for THMs and Cl_3AA through physical adsorption. Replacing the GAC should be staggered and preferably done in early summer, which would better utilize the adsorption capacity of new GAC filters to control higher levels of THMs and Cl_3AA in summer months. After the summer, the saturated and biologically active GAC filters could provide a better biological degradation for HAAs. In a new GAC filter, seeding with acclimated GAC will expedite the biological activity development and could significantly benefit HAA removal.

Longer EBCTs could be used to enhance HAA removal at low water temperatures. In order to achieve a removal efficiency of 50% or higher for HAAs, the authors suggest that a 10-minute EBCT be used for 4°C waters (in the coldest month) and a 5-minute EBCT for waters 10°C or higher.

Biodegradation and organic matrix competition during the isotherm test for HAAs can greatly affect their isotherm constants. The isotherm tests for HAAs should be conducted in the presence of minimal biological activity with the actual filter influent. For HAA removal by physical adsorption, adsorption isotherm data obtained under these conditions provide a better GAC bed life estimation for full-scale GAC filters or contactors.

REFERENCES

Baker, R.A. 1964. Chromatographic Evaluation of Activated Carbon. *Journal AWWA*, 56:92.

Carvalho, M.F., I. Vasconcelos, and A.T. Bull. 2001. A GAC Biofilm Reactor for the Continuous Degradation of 4-Chlorophenol: Treatment Efficiency and Microbial Analysis. *Applied Microbiology Biotechnology*, 57:419–426.

Graese S.L., V.L. Snoeyink, and R.G. Lee. 1987. Granular Activated Carbon Filter-Adsorber Systems. *Journal AWWA*, 79(12):64–74.

Hozalski, R.M., E.J. Bouwer, and S. Goel. 1999. Removal of Natural Organic Matter (NOM) From Drinking Water Supplies by Ozone-Biofiltration. *Water Science and Technology*, 40(9):157–163.

Huck, P.M. 1999. Development of a Framework for Quantifying the Removal of Humic Substances by Biological Filtration. *Water Science and Technology*, 40(9):149–156.

Janssen, D.B., A. Scheper, L. Dijkhuizen, and B. Witholt. 1985. Degradation of Halogenated Aliphatic Compounds by *Xanthobacter autotrophicus* GJ10. *Applied and Environmental Microbiology*, 49(3):673–677.

Jensen, H.L. 1957. Decomposition of Chloro-Substituted Aliphatic Acids by Soil Microorganisms. *Canadian Journal of Microbiology*, 3:151–164.

Keuning, S., D.B. Janssen, and B. Witholt. 1985. Purification and Characterization of Hydrolytic Haloalkane Dehalogenase From *Xanthobacter autotrophicus* GJ10. *Journal of Bacteriology*, 163(2):635–639.

Krasner, S.W., and G. Amy. 1995. Jar-Test Evaluations of Enhanced Coagulation. *Journal AWWA*, 87(10):93–108.

Krasner, S.W., M.J. Sclimenti, and B.M. Coffey. 1992. Biologically Active Filters for the Removal of Aldehydes: An Ozone Pilot-Plant Study. In *Proc. of the AWWA Water Quality Technology Conference*. Denver, Colo.: AWWA.

Lee, C.M., C.J. Lu, and M.S. Chuang. 1994. Effects of Immobilized Cells on the Biodegradation of Chlorinated Phenols. *Water Science and Technology*, 30(9):87–90.

Logan, B.E. 2001. Analysis of Overall Perchlorate Removal Rates in Packed-Bed Bioreactors. *Journal of Environmental Engineering*, 5:469–471.

McGuire, M.J., and I.H. Suffet. 1978. Adsorption of Organics From Domestic Water Supplies. *Journal AWWA*, 70(11):621–636.

Miltner, R.J., and R.S. Summers. 1992. A Pilot-Scale Study of Biological Treatment. In *Proc. of the AWWA Water Quality Technology Conference*. Denver, Colo.: AWWA.

Reckhow, D.A., J.E. Tobiason, M.S. Switzenbaum, R. McEnroe, Y. Xie, Q.-W. Zhu, X. Zhou, P. McLaughlin, and H.J. Dunn. 1992. Control of Disinfection By-Products and AOC by Preozonation and Biologically Active In-line Direct Filtration. In *Proc. of the AWWA Annual Conference and Exposition*. Denver, Colo.: AWWA.

Shi, J., X.D. Zhao, R.F. Hickey, and T.C. Voice. 1995. Role of Adsorption in Granular Activated Carbon-Fluidized Bed Reactors. *Water Environment Research*, 67(3):302–309.

Singer, P.C. 1994. Control of Disinfection By-Products in Drinking Water. *Journal of Environmental Engineering*, 120(4):727–745.

Singer, P.C., H. Arora, E. Dundore, K. Brophy, and H.S. Weinberg. 1999. Control of Haloacetic Acid Concentrations by Biofiltration: A Case Study. In *Proc. of the AWWA Water Quality Technology Conference*. Denver, Colo.: AWWA.

Singer, P.C., A. Obolenski, and A. Greiner. 1995. DBPs in Chlorinated North Carolina Drinking Water. *Journal AWWA*, 87(10):83.

Singer, P.C., R.D. Pyne, M. Avs, C.T. Miller, and C. Mojonnier. 1993. Examining the Impact of Aquifer Storage and Recovery on DBPs. *Journal AWWA*, 85(11):85–94.

Singer, P.C., H.S. Weinberg, K. Brophy, L. Liang, M. Roberts, I. Grisstede, S. Krasner, H. Baribeau, H. Arora, and I. Najm. 2002. *Relative Dominance of Haloacetic Acids and Trihalomethanes in Treated Drinking Water.* Denver, Colo: AwwaRF and AWWA.

Speth, T., and R.J. Miltner. 1990. Technical Note: Adsorption Capacity of GAC for Synthetic Organics. *Journal AWWA*, 82(2):72–75.

Speth, T., and R.J. Miltner. 1998. Technical Note: Adsorption Capacity of GAC for Synthetic Organics. Isotherms of 15 Compounds in Organic-Free Water Are Presented. *Journal AWWA*, 90(4):171–174.

Thomas, J.M., W.A. Mckay, E. Cole, J.E. Landmeyer, and P.M. Bradley. 2000. The Fate of Haloacetic Acids and Trihalomethanes in an Aquifer Storage and Recovery Program, Las Vegas, Nevada. *Ground Water*, 38(4):605–614.

Tung, H.-H., R.F. Unz, and Y.F. Xie. 2002. Mechanisms of Haloacetic Acid Removal by Granular Activated Carbon. In *Proc. of the AWWA Water Quality Technology Conference*. Denver, Colo.: AWWA.

Tung, H.-H., R.F. Unz, and Y.F. Xie. 2003. The Effects of Adsorption Isotherm Testing Conditions on GAC Bed Life Estimation. In *Proc. of the AWWA Annual Conference and Exposition*. Denver, Colo.: AWWA.

Tung, H.-H., M. Yeung, C.A. Cole, and Y.F. Xie. 2003. Using Point-of-Use Carbon Filter to Remove Disinfection Byproducts in Drinking Water. Presented at Innoventure2003, The Penn State University, Hershey, Pa.

USEPA (U.S. Environmental Protection Agency). 1995a. *Method 552.2. Determination of Haloacetic Acids and Dalapon in Drinking Water by Liquid–Liquid Extraction, Derivatization, and Gas Chromatography With Electron Capture Detection*. Cincinnati, Ohio: USEPA, Office of Research and Development.

USEPA (U.S. Environmental Protection Agency). 1995b. *Methods for the Determination of Organic Compounds in Drinking Water*—Supplement III. EPA-600/R-95/131. Washington, D.C.: USEPA, Office of Research and Development.

USEPA (U.S. Environmental Protection Agency). 1998. 40 CFR Parts 9, 141, and 142. National Primary Drinking Water Regulations; Disinfectants and Disinfection Byproducts; Final Rule. *Federal Register*, 63(241):69390–69476.

Weightman, A.L., A.J. Weightman, and J.H. Slater. 1992. Microbial Dehalogenation of Trichloroacetic Acid. *World Journal of Microbiology and Biotechnology*, 8(5):512.

Williams, S.L., D.F. Rindfleisch, and R.L. Williams. 1995. Degradation of Haloacetic Acids at Maximum Residence Time Locations. In *Proc. of the AWWA Water Quality Technology Conference*. Denver, Colo.: AWWA.

Williams, S.L., R.L. Williams, and A.S. Gordon. 1996. The Impact of Bacterial Degradation of Haloacetic Acids in the Distribution System. In *Proc. of the AWWA Water Quality Technology Conference*. Denver, Colo.: AWWA.

Williams S.L., R.L. Williams, and A.S. Gordon. 1997. Degradation of Haloacetic Acids in Drinking Water. In *Proc. of the 213th ACS National Meeting, Division of Environmental Chemistry*. San Francisco, Calif.: American Chemical Society.

Williams, S.L., R.L. Williams, and D.F. Rindfleisch. 1994. Dead End on Haloacetic Acids. In *Proc. of the AWWA Water Quality Technology Conference*. Denver, Colo.: AWWA.

Williams, S.L, R.L. Williams, and J. Yuan. 1998. Bacterial Degradation of Haloacetic Acids in the Distribution System. In *Proc. of the AWWA Water Quality Technology Conference*. Denver, Colo.: AWWA.

Wobma, P., D. Pernitsky, B. Bellamy, K. Kjartanson, and K. Sears. 2000. Biological Filtration for Ozone and Chlorine DBP Removal. *Ozone: Science and Engineering*, 22(4):393–413.

Wu, H., and Y.F. Xie. 2003. Effects of Empty Bed Contact Time and Temperature on the Removal of Haloacetic Acids Using Biologically Active Carbon. In *Proc. of the AWWA Annual Conference and Exposition*. Denver, Colo.: AWWA.

Xie, Y.F. 2004. *Disinfection Byproducts in Drinking Water: Formation, Analysis, and Control.* Boca Raton, Fla.: Lewis Publishers.

Xie, Y.F., I. Rashid, H. Zhou, and L. Gammie. 2002. Acidic Methanol Methylation for HAA Analysis: Limitations and Possible Solutions. *Journal AWWA*, 94(11):115–122.

Xie, Y.F., and D.A. Reckhow. 1992. A New Class of Ozonation Byproducts: The Ketoacids. In *Proc. of the AWWA Annual Conference and Exposition.* Denver, Colo.: AWWA.

Xie, Y.F., D.A. Reckhow, and D.C. Springborg. 1998. Analyzing HAAs and Ketoacids Without Diazomethane. *Journal AWWA*, 90(4):131–138.

Xie, Y.F., Wu, H., and H.-H. Tung. 2003. GAC for DBP Removal: When and How to Change GAC? In *Proc. of the AWWA Water Quality Technology Conference*. Denver, Colo.: AWWA.

Xie, Y.F., and H. Zhou. 2002. Use of BAC for HAA Removal—Part 2, Column Study. *Journal AWWA*, 94(5):126–134.

Zhang, X.J., L.L. Cao, Z.S. Wang, W.J. Liu, and S. Li. 1998. Adsorption Characteristics of Haloacetic Acids by Activated Carbon in Drinking Water Treatment. In *Water and Wastewater Industry and Sustainable Development.* Edited by Y. Qian, J.M. Hao, X.J. Zhang, and H.T. Zhang. Beijing, China: Tsinghua University.

Zhou, H., and Y.F. Xie. 2002. Using BAC for HAA Removal—Part 1, Batch Study. *Journal AWWA*, 94(4):194–200.

ABBREVIATIONS

AWWA	American Water Works Association
AwwaRF	Awwa Research Foundation
BAC	biologically active carbon
BL	bed life
CIA	capillary ion analyzer
cm	centimeter
DBP	disinfection by-product
D/DBP	disinfectants/disinfection by-products
DO	dissolved oxygen
EBCT	empty-bed contact time
FI	filter influent
ft^2	square foot
g	gram
GAC	granular activated carbon
GC–ECD	gas chromatography–electron capture detector
gpm	gallons per minute
HAA	haloacetic acid
HAA5	the sum of five haloacetic acids: monochloroacetic acid (ClAA), dichloroacetic acid (Cl_2AA), trichloroacetic acid (Cl_3AA), monobromoacetic acid (BrAA), and dibromoacetic acid (Br_2AA)
HAA6	the sum of six haloacetic acids: monochloroacetic acid (ClAA), dichloroacetic acid (Cl_2AA), trichloroacetic acid (Cl_3AA), monobromoacetic acid (BrAA), dibromoacetic acid (Br_2AA), and bromochloroacetic acid (BrClAA)
hr	hour
ID	inside diameter
in.	inch
L	liter
LT	literature
m	meter
MCL	maximum contaminant level
mg	milligram
min	minute
mL	milliliter

mm	millimeter
mM	millimolar
MTBE	methyl-*tert*-butyl-ether
μA	microampere
μg	microgram
μm	micrometer
μM	micromolar
NA	not available; not applicable
nm	nanometer
NOM	natural organic matter
OFW	organic-free water
PTFE	polytetrafluoroethylene
rpm	revolutions per minute
THM	trihalomethane
THM4	the sum of four trihalomethanes: chloroform ($CHCl_3$), bromoform ($CHBr_3$), dichlorobromomethane ($CHCl_2Br$), and chlorodibromomethane ($CHBr_2Cl$)
U.S.	United States
USEPA	United States Environmental Protection Agency
UV	ultraviolet

IWA Publishing's authorised EU representative for General Product Safety Regulations is Diane D'Arras, 15 rue Duret, 75116 Paris, France, e-mail: safety@iwap.co.uk.

Printed and bound by CPI Group (UK) Ltd, Croydon, CR0 4YY
17/09/2026
02218811-0007